Hidden Minds

A History of the Unconscious

Frank Tallis

Arcade Publishing • New York

FIRST NORTH AMERICAN EDITION

First published in 2002 by Profile Books Ltd in Great Britian

Library of Congress Cataloging-in-Publication Data

    Tallis, Frank.
          Hidden minds : a history of the unconscious / Frank Tallis.
            p. cm.
          Includes bibliographical references and index.
          ISBN 1-55970-643-0
          1. Subconsciousness. 2. Subconsciousness—History. I. Title.

          BF315 .T32 2002b
          154.2—dc21   2002074566

Published in the United States by Arcade Publishing, Inc., New York
Distributed by AOL Time Warner Book Group

Visit our Web site at www.arcadepub.com

10 9 8 7 6 5 4 3 2 1

EB

PRINTED IN THE UNITED STATES OF AMERICA

To Nicola

Contents

Acknowledgements

I would like to thank the following: Professor Michael J. Power for reading the first draft of this book and providing many helpful comments, Andrew Franklin for his expert editorial guidance, Wyllys Poynton for locating Myers' (1886) article on the work of Pierre Janet, and Dr Jackie Andrade for directing me towards relevant research on learning under general anaesthesia. I would also like to thank Dr Doris Silverman and Dr Carolyn Ellman for providing me with biographical information on the late Lloyd Silverman. Finally, I would like to thank Nicola Fox for her invaluable assistance in the production of this book – at every stage, from conception to final draft. Needless to say, any flaws, errors, or shortcomings are entirely mine.

Introduction

Perhaps the first individual to acknowledge that some parts of the mind are necessarily unavailable for introspection was St Augustine. He wrote 'I cannot grasp all that I am', meaning that at any single point in time he could be aware of only a fraction of his totality. All his memories and knowledge – most of what contributed to his sense of self – remained beyond awareness. Augustine recognised that consciousness has a limited capacity. It is only possible to suspend a small number of images or words in the medium of consciousness. Yet, even when our conscious minds are fully occupied – watching a sunset, or reading a book – the 'sense' of being who we are is generally well preserved. Our personal history – recorded in the form of memories – although far too extensive to be squeezed into a single moment of awareness, continues to impress on us a unique sense of identity. We feel the presence of our unconscious mind like a ghost. Invisible, but nevertheless somehow there.

The concept of the unconscious is surprisingly ubiquitous; scientists and poets, philosophers and artists, neurologists and mystics have all explored or discussed the concept of the 'unconscious'. It has been the subject of learned dissertations as well as films, plays, and even a hit Broadway musical. But what is it, exactly? And how important is the unconscious in contemporary accounts of the human mind?

Although the unconscious has been approached from a variety of perspectives, the term unconscious is generally employed to refer to parts of the mind (or processes operating within the mind) that are either permanently or temporarily inaccessible to awareness; however, beyond this simple description, two contrasting ideas of the unconscious have repeatedly found (or lost) favour.

The first imbues the unconscious with many of the properties associated with consciousness: it can analyse information, make judgements, and sanction decisions. As such, it resembles a kind of hidden auxiliary intelligence. The second idea invests the unconscious with the properties of a machine. Information in the unconscious is simply processed by the neurological equivalent of a factory production line – disinterested, unreflecting, and automatic.

Artists and writers have always favoured the former view. There are numerous accounts of great works of art being produced without any conscious effort. Schiller, Goethe, Mozart, Coleridge, and Blake all produced masterpieces which seemed to enter awareness whole – completed and ready for transcription; however, creative artists are not the exclusive beneficiaries of a bountiful unconscious. Many scientists and mathematicians – for example Poincaré, Gauss, Kekule, Bohr, and Einstein – also arrived at the solution to complex problems while dozing or simply going about their usual everyday tasks. Revelatory experiences suggest the existence of a sophisticated unconscious, capable of undertaking complex mental operations. Clearly, such an unconscious is a valuable resource. But can the unconscious be exploited? Can it be encouraged to produce great works of art and solve problems? And if instructed, will it obey?

In the past, many methods have been employed to tap the unconscious and harness its powers. In the eighteenth century early hypnotists discovered that the unconscious could be 'programmed' to influence behaviour long after subjects were roused from their hypnotic sleep. In the nineteenth century poets and writers took opium in order to liberate the contents of the unconscious (and were rewarded with awe-inspiring visions and dreams). However, the most recent and scientifically supported method of communicating with the unconscious is subliminal stimulation. This technique conventionally involves the very short presentation (for five milliseconds or less) of messages or instructions. These presentations are so brief that they are not registered consciously, but nevertheless seem to enter and bias the processing systems of the brain. Some studies suggest that subliminal messages influence thoughts, feelings, and behaviour more significantly than the same messages shown above the awareness threshold.

Subliminal stimulation studies suggest that the unconscious can be persuaded to influence the conscious mind. But to what extent does the unconscious influence us ordinarily? How important is activity in the unconscious? We have the subjective impression that the conscious mind is very much the dominant partner. Yet this impression might be inaccurate and misleading. This was certainly the belief of Sigmund Freud, the man whose name is now most frequently linked with the concept of the unconscious.

In his *Introductory Lectures on Psychoanalysis*, written between 1915 and 1917, Sigmund Freud claimed to have delivered 'the third blow' to narcissistic humanity. Copernicus, he said, had delivered the first – wrenching the earth away from the universe's centre – and Darwin the second, showing man's descent from ape-like ancestors. By emphasising the importance of unconscious processes in mental life, Freud believed that he had delivered the third and most wounding blow in the sequence. He suggested that our most valued characteristics – free will, rationality and a sense of self – are mere illusions, and that we are all the products of unconscious and uncontrollable forces in the mind. Naturally, Freud met with considerable resistance.

Many commentators dismissed Freud's claim as grandiose and by the middle of the twentieth century Freud and his ideas were falling out of favour. Behaviourism (a theoretical framework which eschewed consideration of mental states altogether) held sway in academic psychology, while in the clinic advances in drug treatments threatened to make psychoanalysis redundant. In addition, the advent of the modern computer had provided psychology with a powerful new framework and vocabulary with which to understand and describe mental processes. Subsequently, the unconscious – the mainstay of Freud's intellectual heritage – seemed less relevant. It seemed like an idea too strongly associated with the past to have much of a future.

But what if Freud had been right? What if Freud did deliver the third blow? And what if the concept of the unconscious is every bit as important – and humbling – as the heliocentric universe or evolution? We are all familiar with the massive cultural impact of Copernicus and Darwin, but how would humanity respond to the final assault on its egocentricity? How would these new insights affect issues surrounding the exercise of choice and freewill? Issues surrounding the nature of identity? Selfhood? Morality?

Over the past fifty years the unconscious has made a comeback. Evolutionary theorists, neuroscientists, experimental psychologists, and those working in the field of artificial intelligence have all been forced to reconsider the concept of the unconscious. Gradually, individuals working in different disciplines, on different problems, but linked by a common goal of wishing to elucidate the properties of the mind, have been giving more and more emphasis to unconscious mental processes. Indeed, in the past half century the fortunes of the unconscious have been completely reversed. It is now almost impossible to construct a credible model of the mind without assuming that important functions will be performed outside of awareness.

One of the greatest misconceptions about Freud is that he 'discovered' the unconscious. He didn't. In fact, most of Freud's ideas about the unconscious had been delineated and discussed (largely by philosophers and physicians) for over a century before his first relevant publications. Nevertheless, Freud can be justly credited with recognising that the unconscious was a very significant idea, and then popularising it with almost fanatical enthusiasm. Unlike his predecessors, Freud understood that the existence of the unconscious necessitated a complete revision of how human beings saw themselves. He rescued the unconscious from the rarefied world of academic debate and made it an indispensable addition to the modern vocabulary. Although Freud wrote extensively on many aspects of mental life, he felt that his greatest achievement was recognising the importance of the unconscious – hence, his unmistakable pride in having delivered 'the third blow'.

Important ideas take a long time to yield up all their implications. In the custody of each new generation, a further application is discovered – another overtone, another resonance. Moreover, important ideas propagate slowly. There is usually a substantial delay between the inception of an idea and its subsequent widespread appreciation (particularly among the general public).

Although Copernicus proposed the heliocentric (sun-centred) universe in the sixteenth century, it wasn't until the seventeenth or even eighteenth century that the educated classes of Europe realised that his observations represented much more than an astronomical discovery. The human race had been demoted. The human race no longer occupied a pre-eminent position in the cosmos. After Copernicus, the old medieval certainties (including an assumed existence of God) seemed less robust.

Likewise, it took over a hundred years for Darwin's theory of evolution to be fully appreciated. Although evolution was immediately perceived as a challenge to religious orthodoxy (human beings were not created, but shaped by natural selection), it wasn't until the late twentieth century that the awesome explanatory power of Darwin's idea was really understood. Evolution could account for everything from the behaviour of micro-organisms to human altruism, the dance of the honey bee to men in fast cars. Moreover, Darwin's principles had applications well beyond the confines of natural history.

In the early years of the twentieth century Freud asserted that almost all of the processes which determine who we are, and what we do, occur below the awareness threshold. Following the pattern set by his illustrious predecessors Copernicus and Darwin, it has taken nearly a century for the full significance of Freud's ideas about the unconscious to be recognised. In the academic scramble to discredit Freud – particularly some of his more outlandish ideas on sex and sexual development – this, his most fundamental and defining statement, was never given the proper consideration it deserved. It is only now that the deeper implications of Freud's 'third blow' are being fully appreciated.

Until relatively recently neuroscientists were in the habit of postulating some sort of 'central executive' in the brain. A place where consciousness 'happened'. This idea is not very different from one proposed by Descartes, who in the seventeenth century suggested that consciousness was synthesised in the pineal gland. New models of brain functioning have, however, completely abandoned this approach.

Brain activity is distributed among numerous specialised sub-systems (i.e. groups of cells that work together to perform a particular function); yet none of these sub-systems is conscious. Moreover, there is no special place in the brain where the products of these sub-systems (such as the objects of perception and emotions) are knitted together. No place where a conscious experience is constructed. This has led many neuroscientists to conclude that the sense of self that we all share is nothing more than an 'illusion'; that *identity* is to the brain what the shape of a wave is to sea water. Contemporary neuroscience seems to suggest that our most prized possession – the ego – barely exists. We (as we know ourselves) have only a feeble claim on existence.

These are exciting times. With the advent of brain-scanning technology it is now possible to map the brain. Already the geography of the unconscious has been illuminated by the fallout of colliding sub-atomic particles. Many brain scan images – showing brightly lit areas of biological activity – are nothing less than snapshots of the unconscious at work. Preconscious processes, rapidly assembling the infrastructure of personality.

The unconscious, only recently rejected as a historical curiosity, has made its way back to the heart of neuroscience. It is now widely recognised that without a thorough understanding of unconscious processes in the brain we will never have a thorough understanding of ourselves. Once again, the unconscious is an idea with a future – but to appreciate that future, we must first consider its past.

Hidden Minds

1

Depths below depths

When the German philosopher Gottfried Wilhelm Leibniz discovered John Locke's *Essay Concerning Human Understanding*, he was greatly impressed. This work, published in 1690, was a meticulous analysis of how knowledge becomes consolidated in the mind. Although excited by Locke's 'essay', Leibniz harboured some reservations and subsequently penned a brief critique. Unfortunately, Locke took a very dim view of Leibniz's personal communication and chose to ignore it. Still convinced that his critique was sound, Leibniz wondered if Locke might have simply misunderstood what he had written. So, in the interests of clarity, he expanded his arguments to fill a book – *New Essays on Human Understanding*. But just as Leibniz completed it, Locke died. Disinclined to challenge an adversary who suffered from the considerable handicap of being dead, Leibniz sportingly decided not to publish; however, his *New Essays* did finally appear in 1765, almost fifty years after his own death.

New Essays is of major importance, not only because it contains a well-argued response to Locke but also because it contains the first significant entry into philosophical discussion of unconscious mental operations. Although others (such as St Augustine, Montaigne, and Descartes) had speculated about inaccessible memories or actions undertaken in the absence of awareness, never before had unconscious processes received such detailed consideration.

Leibniz's new way of understanding the mind, as a marriage of conscious and unconscious parts, initiated a tradition that eventually influenced the entire development of German psychology (up to and including Freud); however, the significance of Leibniz's work was not fully appreciated when *New Essays* was published. Indeed, proper recognition would be delayed for almost a century. This was because Leibniz's book appeared when the Age of Enlightenment was approaching its zenith, and during the Enlightenment the power of reason had been given considerable emphasis in all accounts of mental life. Leibniz's proposal, that there might be unconscious processes at work in the mind affecting the formation of ideas, judgement, and decision making, was in stark contradiction to the prevailing view. In an age that respected mastery and control, Leibniz's ideas were both unwelcome and perhaps more than a little disconcerting. It was absurd, surely, to suggest that man (equipped with 'god-like reason') should be influenced by mental events so insubstantial as to escape his ordinary notice. Or so it was thought.

Whether considering the motion of gases or the orbit of Mr Halley's comet, reason was showing that all phenomena were lawful. Even the passage of time itself could be measured with greater accuracy. Throughout the seventeenth and eighteenth centuries table clocks and long clocks began to appear in the drawing rooms of Europe. These clever devices, with their springs and bobs, click wheels and ratchets, provided the age with a guiding metaphor. The universe was like an enormous clock. The universe ran like clockwork. The idea that human society and the human mind itself might run on similar principles was welcomed as yet another pleasing symmetry.

Locke's *Essay* with which Leibniz was to take posthumous issue contains all of the hallmarks of Enlightenment thinking: complex mental phenomena are broken down into more fundamental constituents and 'opposition to reason' is described as 'a sort of madness'. Moreover, in the fourth edition Locke began to explore potential laws which might determine how certain ideas (e.g. size and shape) become associated. Locke believed that self-reflection could unravel the mind completely. Although he recognised that ideas might exist outside of awareness – stored in memory – such ideas could easily be uncovered at will. There were no inaccessible recesses or shadowy corners. The machinery of mental life could be exposed through introspection, just as one might peer into the back of a table clock to examine its cogs and springs. This was the key difference between Locke and Leibniz. For Locke the mind was transparent, but for Leibniz the mind was semi-opaque.

Leibniz was, without doubt, an extraordinary individual: uncommonly gifted and enjoying an embarrassment of intellectual riches. He is perhaps most famous for discovering the calculus – a systematic method of calculating areas, volumes, and other quantities very much superior to anything that existed before. Unfortunately, Newton was working on the same problem at exactly the same time, resulting in an acrimonious priority dispute. Nevertheless, it was Leibniz's method of calculation that proved to be less cumbersome, giving continental mathematicians a significant advantage over their loyal but misguided English counterparts. In addition to discovering the calculus, Leibniz also devised binary arithmetic – a means of representing all values with only ones and zeros and better known today as the 'language' used by all digital computers. Indeed, it can be argued that Leibniz anticipated computer science itself by inventing a machine (the wooden prototype of which he demonstrated in London in 1673) capable of 'reasoning' by manipulating a symbolic language (over 150 years before Babbage's Analytical Engine). As if this wasn't enough, he designed a submarine, anticipated some features of Einstein's theory of relativity, improved some basic engineering designs, promoted a public health system (which included a firefighting service and street lighting), assisted in negotiations that secured Georg Ludwig of Hanover's succession to the British throne, and helped establish the German State Bank. He was no slacker.

Be that as it may, when John Locke read Leibniz's preliminary critique of his *An Essay Concerning Human Understanding* he allegedly responded by saying:

> Mr L's great name had raised in me an expectation which the sight of his
> paper did not answer. This sort of fiddling makes me hardly avoid thinking
> that he is not the very great man as has been talked of him.

Although Leibniz was undeniably an inveterate 'fiddler', there can be no denying that his fiddling was of the highest quality. Moreover, when he finally decided to 'fiddle' with mental phenomena, it was inevitable that his mighty intellect would deliver a revolutionary and penetrating account of the mind.

Leibniz did not make a sharp distinction between awareness and its absence. He believed that even when the mind is ostensibly inactive, such as in dreamless sleep, at some level mental processes are still operating. In addition, he postulated a continuum of consciousness. At the top of this continuum, he placed *apperceptions* (that is, the occurrence of clear and distinct mental experiences). Below these were less well defined *perceptions*, and below these *minute perceptions*, which occur wholly outside of awareness because they are 'either too minute and too numerous, or else too unvarying'; however, minute perceptions do not always remain unconscious. They can rise into awareness, as when an individual focuses attention on a previously unnoticed sensation or noise.

Leibniz attributed a central role to minute perceptions with respect to the creation and maintenance of our sense of identity. According to Leibniz, the sense of having a single, continuous identity, extending from childhood through to old age, arises because of a sub-stratum of unconscious memories. Some of these memories enter awareness completely (allowing the individual to cross reference his or her past), while other memories merely hover at the fringes of awareness, providing a tenebrous context against which current mental events can take place. Leibniz also suggested something else that must have been particularly unpalatable to his peers; something which we today would call *unconscious motivation*. Again, minute perceptions played a key role. Leibniz suggested that minute perceptions might influence choices (and subsequent behaviour) without ever being detected in awareness. In other words, human beings were creatures with limited self-knowledge or insight. They might not be wholly aware why they choose to act in one way rather than another. Understandably, the concept of unconscious motivation never gained much currency in an age where introspection was thought to reveal the ordered workings of a wholly transparent and rational mind. The clear implication was that human beings were fundamentally irrational – a ridiculous idea. Unthinkable.

It is ironic that Leibniz employed a mechanical image to explain how minute perceptions influence behaviour, thus inadvertently subverting the most potent symbol of the Enlightenment. For Leibniz, minute perceptions resemble 'so many little springs trying to unwind and driving our machine along'. He goes on to explain:

> That is why we are never indifferent, even when we appear to be most so, as for instance whether to turn left or right at the end of a lane. For the choice that we make arises from these insensible stimuli, which, mingled with the actions of objects and our bodily interiors, make us find one direction of movement more comfortable than the other.

Leibniz's response to Locke (with its emphasis on unconscious determinants of behaviour) was a controversial publication in the Age of Reason. In addition, Leibniz's reputation may have suffered when the French writer and wit, François Marie Arouet, more famously known as Voltaire, ridiculed him as the absurd philosopher Dr Pangloss in *Candide* (1759). Subsequently, Leibniz's revolutionary ideas about the workings of the mind were somewhat neglected.

Even so, as the well-oiled wheels of the table clocks turned, marking time with increasing precision, the world was edging forward to meet a new century; a new century in which mechanistic models of mind would be rejected and the concept of the unconscious fully embraced. 'Mock on, Mock on, Voltaire', wrote the poet William Blake, ''tis all in vain!' And he was right. The tectonic plates of art and philosophy suddenly shifted. Within a few decades the citadels of reason would be reduced to rubble.

The romantic movement began in Germany towards the end of the eighteenth century; however, the romantic sensibility continued to be influential to the end of the nineteenth, affecting cultural life worldwide. Although conventionally considered an artistic movement, the influence of romanticism spread well beyond the arts. Indeed, the representatives of romanticism offered a new model of man and, inevitably, a new model of mental life. Moreover, right from the very beginning, the existence of the unconscious was fully accepted and integrated into romantic psychology. The poet Friedrich Schiller suggested that his poetry had an unconscious origin and argued that the creative faculties were improved when liberated from reason; Johann Wolfgang von Goethe claimed to have written his influential *The Sorrows of the Young Werther* (1774) while 'practically unconscious'; and philosophers such as Artur Schopenhauer began to describe man as an irrational creature driven by unconscious forces. Why the sudden change?

Romanticism was very much a reaction against the central values of the Enlightenment. Indeed, the principal themes of romanticism were established in the wake of a philosophical backlash. The Enlightenment had arisen in the capitals of Europe, where man had been defined as a social and rational animal. Subsequently the romantics exchanged the city for the countryside, society for solitude, and reason for emotion.

The romantics venerated nature, and this is clearly evident in early eighteenth century romantic art – the most typical examples of which were produced by painters such as Caspar David Friedrich. Friedrich (and his contemporaries) were preoccupied with nature's power and majesty, and subsequently specialised in landscapes. Human beings occasionally appear, but they are usually depicted as insignificant – dwarfed by immense mountains and louring skies, huge cataracts or sheets of ice. Solitary figures look out over vast expanses of rolling mist, or huddle on beaches in moonlight. Many of the greatest triumphs of the Enlightenment were urban, architectural (Sir Christopher Wren's St Paul's Cathedral, for example). The raw materials of nature – slate, marble, granite – had been disciplined by mathematics and labour; however, it is difficult to imagine Friedrich's proud rocks being shaped into Corinthian columns or decorative cornicing. For Friedrich and the romantic fraternity nature was wild and elemental. It represented something beyond the scope of reason, ratio and compasses.

The solitary artists, rejected lovers, and lonely wayfarers in romantic landscapes also reflect the special emphasis romanticism gave to the individual. Enlightenment thinkers had wrestled with political and economic issues – analysing social structures and planning social reforms. With the advent of romanticism, the individual became the focus of interest. Writers and artists demonstrated an increasing awareness of the complexity of mental life. The struggles between different elements of society were now of less concern than the struggles arising within the individual – the conflict between head and heart, body and soul, conscious and unconscious forces.

As romanticism gathered momentum, the faculty of reason was approached with less reverence, which permitted more serious consideration of the subjective, irrational, and visionary. Deep feelings, or passions, previously viewed with some suspicion (on account of being wrongly associated with mental illness) were increasingly perceived as desirable and enriching. Even mysticism (rejected by Enlightenment thinkers as superstitious nonsense) became more acceptable. Indeed, one of the pivotal beliefs of romanticism was that behind visible nature was a mysterious 'fundament' or *Grund*. Subsequently, the romantic movement's professed love of nature concealed an ulterior motive: union with a kind of universal unconscious – the cosmic equivalent of the soul's penetralia.

According to romantic philosophers, the universal unconscious contained its own memories. It was a storehouse of ancient lore, symbols, and leitmotif. Because the human unconscious resonated in sympathy with its deep, immemorial voice, certain themes and images were prone to recur in myth and folklore – all issuing from the same source. Needless to say, union with the universal unconscious was accomplished most successfully when the faculty of reason was suspended. Thus, powerful emotions, mental illness, inspiration, and dreams acquired special significance. Such altered states punctured the membrane separating lacklustre reality from the numinous, allowing primordial knowledge and impressions to seep into awareness. These views are evident in works such as *The Symbolism of Dreams* (1814) by Gotthilf Heinrich von Schubert, who argued that dreams incorporate universal, timeless symbols that can be interpreted irrespective of their temporal and geographical provenance.

By the early decades of the nineteenth century the unconscious had evolved into something very far removed from Leibniz's now quite humble-sounding 'minute perceptions'. It had acquired a complex set of properties and associations reflecting the preoccupations of romanticism. The unconscious was the font of inspiration and creativity – a hidden anvil on which the products of imagination were mysteriously hammered into existence. A dream factory in which the elements of arcane symbols were welded together. The source of deep feelings and a portal through which one might glimpse universal truths. In addition, a further sense of the unconscious had begun to emerge: that of the unconscious as a place or location. Somewhere inside the head that could actually be visited. A place where the Newtonian prison of time and space could be escaped from. A place of endless possibilities.

The notion of descent into a shadowy kingdom (albeit a descent that passes through layers of mind) resonates strongly with some of the most enduring myths of the western literary canon: Orpheus, Aeneas, and Dante all descended into their respective underworlds. It was inevitable that, once the unconscious was conceptualised as a place, people would want to go there. And in due course they did. But many of these early mind-travellers required a little help – from *Papaver somniferum*, the opium poppy. Nature's passport to the unconscious.

The birth of romanticism in English literature is associated with the publication of the *Lyrical Ballads* (1798) by William Wordsworth and Samuel Taylor Coleridge. In this and subsequent work it is plain that both men believed that the human mind was far more complex than a table clock, and its working infinitely more mysterious. Coleridge wrote of 'the twilight realms of consciousness', while Wordsworth described 'Caverns ... within my mind which sun/ Could never penetrate'.

It has been argued that the fundamental motivation for romantic creativity was not self-expression but self-exploration, and indeed many of the poems of Wordsworth and Coleridge are perhaps best considered as products of self-enquiry or self-analysis. Of these two poetic giants, the process of self-enquiry is ostensibly more obvious in the work of Wordsworth. His epic poem *The Prelude, or, Growth of a Poet's Mind* (1850), took nearly forty years to write and describes the development of Wordsworth's own imagination. Although *The Prelude* is a verse autobiography, it is unusual (compared with previous autobiography) insofar as greater emphasis is given to internal, mental events than events that happen 'out there' in the 'real world'.

Coleridge never wrote a vast autobiographical poem that compares with *The Prelude*; however, he was just as interested in his own mental processes – possibly even more so than Wordsworth was. Coleridge wrote extensively on the mind and was always jotting down relevant observations. Like Wordsworth, he had a special interest in the imagination, but he also wrote a great deal on topics that we would now recognise as 'psychology' (such as perception, memory, and melancholia). Indeed, Coleridge is often credited with having introduced the word 'psychology' into the English language (albeit as a German import).

Coleridge's reflections on the mind and the imaginative faculty are arguably more penetrating than Wordsworth's, but this may be because he had an unfair advantage. Unlike his more sober friend, Coleridge was into drugs. In the 1800s, opium was readily available from pharmacists, druggists, and physicians, who recommended its use as an analgesic. It was conventionally taken as laudanum, a tincture of opium diluted in wine or brandy. By 1801 Coleridge was addicted, and such were his experiences under the influence of opium that he speculated on the nature of mental phenomena for the rest of his life.

In 1816 – at the suggestion of a young Lord Byron – Coleridge published his unfinished 'Kubla Khan', a poem set in a strange, exotic landscape, replete with images of underground rivers, 'caverns measureless to man', and sunless seas. Coleridge consented to the publication of 'Kubla Khan', as a 'psychological curiosity', rather than a work of great artistic merit. This was largely due to the curious circumstances surrounding its composition.

Almost twenty years earlier Coleridge had been living in a 'lonely farmhouse' between Lynton and Porlock on the edge of Exmoor. He had taken some opium for a 'medical complaint', and while asleep experienced the spontaneous composition of an enormous, visionary poem, beginning with the now famous lines 'In Xanadu did Kubla Khan/ A stately pleasure dome decree.' After writing a page or so Coleridge was disturbed by 'a person on business from Porlock', and when Coleridge returned to complete the poem he was unable to do so. It was as though an internal door had closed. He no longer had access to his poetic vision.

'Kubla Khan' is extraordinary for many reasons. First of all, it is, as Coleridge suggested, a psychological curiosity. Its method of composition is as fascinating as its content. Coleridge felt that he could not properly claim authorship of 'Kubla Khan', as it was produced without 'conscious effort'. It had bubbled up, just like one of the underground streams described in the poem, from his unconscious. Its very existence contradicted the fundamental assumptions made in the eighteenth century concerning the nature of poetic creativity. Poems were supposed to be crafted. Poems were the result of skill, artifice. They weren't supposed to rise into awareness, complete, while the 'author' was asleep. Secondly, Coleridge's sparkling description of Xanadu is so vivid it's almost as though he went there. 'Kubla Khan' is nothing less than a postcard from Xanadu. And where was Xanadu? It could only be in one place – Coleridge's own mind. But a part of his mind only accessible during sleep. The deep, fathomless sleep of an opium addict.

Unfortunately for Coleridge, once he had become addicted to opium the pleasure domes of Xanadu were swiftly replaced by a sort of bespoke version of hell. Coleridge's notebooks are full of references to either disturbing opium-related experiences or the dreadful physical symptoms that accompanied periods of abstinence. Nights were particularly bad. He wrote of 'these Sleeps, these Horrors, these Frightful Dreams of Despair' and he was visited by 'a pandemonium of all the shames and miseries of the past ... bronzed with one stormy Light of Terror & Self-Torture.' He also suffered from hallucinations. Opium seemed to liberate the contents of the unconscious. The wellhead of imagination could be chemically encouraged to release its enchanted rivers, colourful demons, and unwanted memories into the narrative of dreams and hypnagogic visions.

Although Coleridge became well acquainted with his own unconscious (and its products), he never published a detailed account of his travels there. This task was undertaken by Coleridge's young admirer, Thomas De Quincey, whose scandalous *Confessions of an English Opium Eater* (1822) is arguably the first comprehensive literary celebration of unconscious mental life.

De Quincey was only twenty-one when he first met Coleridge in 1807. Just before this first meeting he had been told by a mutual friend that latterly Coleridge was less concerned with poetry and instead 'applied his whole mind to metaphysics and psychology'. The accuracy of this assessment appears to be have been borne out by the subsequent conversation, which was rather one-sided and consisted mostly of an extended philosophical monologue courtesy of the older man. Although De Quincey's first impression of Coleridge was of a corpulent, dreamy individual, Coleridge soon surprised De Quincey with his eloquence and ability casually to offer solutions to philosophical problems 'that had existed for twenty-three centuries.'

De Quincey wasn't, however, terribly impressed by Mrs Coleridge. But then, neither was Coleridge. While the two men were talking a woman appeared whom De Quincey described rather uncharitably as 'full and rather below the common height' (somewhat rich, given contemporary accounts of De Quincey's own diminutive stature). Coleridge turned to De Quincey and made an introduction which was as economic as it was frigid. 'Mrs Coleridge,' he said. Whereupon Mrs Coleridge quickly withdrew. 'I gathered', wrote De Quincey, '… that Coleridge's marriage had not been a very happy one.'

That evening De Quincey remarked to Coleridge that he had been obliged to take a few drops of laudanum because of a toothache, at which point Coleridge warned De Quincey very strongly 'against forming a habit'. Fortunately for English literature De Quincey took absolutely no notice of Coleridge's advice. Although – in actual fact – this particular piece of advice was already far too late.

In the autumn of 1804 De Quincey had suffered from what he described as 'rheumatic pains of the head and face'. These pains had lasted for some twenty days. On the twenty-first day – a dreary, wet Sunday – De Quincey got up and, with no particular purpose in mind, wandered the streets of London. By chance he encountered an acquaintance who, on hearing about De Quincey's affliction, urged him to try opium as a remedy. De Quincey immediately went shopping on Oxford Street, and was soon able to obtain a tincture of opium from the local 'druggist'.

When De Quincey got back to his lodgings he wasted no time in taking the prescribed quantity. His response is best described in his own words:

> Here was a panacea … for all human woes: here was the secret of happiness,
> about which philosophers had disputed for so many ages, at once discovered:
> happiness might now be bought for a penny, and carried in the waistcoat
> pocket: portable ecstacies might be had corked in a pint bottle: and peace of
> mind could be sent down in gallons by the mail coach.

Clearly, the effect of opium had been considerably more far-reaching than facial anaesthesia. De Quincey had had quite an experience. Yet, surprisingly (given De Quincey's encomium), he didn't become addicted immediately. Although he took opium regularly, it wasn't until 1813 that he became dependent. That year marked the return of an old stomach complaint which De Quincey tried to manage by increasing his opium intake. The dose escalated, his use became immoderate, and thereafter he referred to himself as a 'confirmed opium-eater'.

Like Coleridge, De Quincey began to hallucinate. He would lie awake, watching incredible scenes unfolding at the end of his bed, 'vast processions' and 'friezes of never ending stories'; however, these processions and friezes paled into insignificance when compared with his subsequent dreams – 'a theatre', he wrote, 'seemed suddenly opened and lighted up within my brain, which presented nightly spectacles of more than earthly splendour'.

The realm that De Quincey visited in his dreams was not an elevated, transcendent realm. His consciousness did not rise up to occupy a vantage point in a higher, more refined reality. Rather, De Quincey had a distinct impression that he was sinking into himself:

> I seemed every night to descend, not metaphorically, but literally to descend,
> into chasms and sunless abysses, depths below depths, from which it seemed
> hopeless that I could ever reascend.

Unlike Coleridge, who peered into the chasm of the unconscious with a certain amount of dread and horror, one can't help feeling that De Quincey undertook his descent with unnatural glee. For someone supposedly writing about the pains of opium, his language is suspiciously rapturous. He seems to relish the experience. Images conjured before sleep were 'drawn out by the fierce chemistry' of dreams 'into insufferable splendour'; he saw 'gorgeous spectacles'.

De Quincey observed that in the dream world the rules of time and space (those stalwarts of the Newtonian universe) broke down completely: 'Space swelled, and was amplified to an extent of unutterable infinity.' Thus, De Quincey describes walking in impossible buildings, whose colossal architecture was well beyond the scope of Christopher Wren's mathematics. Time became entirely meaningless. 'I sometimes seemed', he wrote, 'to have lived for 70 or 100 years in one night; nay, sometimes had feelings representative of a millennium passed in that time ...' He was trapped in pagodas for centuries, walked through narrow chambers in 'eternal pyramids', assumed different identities – from sacrificial victim to high priest – spoke with gods and animals, kissed crocodiles, and lay down in the mud of the Nile.

Confessions of an English Opium Eater acquired cult status almost immediately after publication; however, De Quincey also wrote a lesser known sequel, *Suspiria de Profundis* (1845), which is in many ways even more fascinating. It is an extremely odd, lyrical work – a piece of preemptive psychedelia – populated by a cast of characters that, again, De Quincey became acquainted with during the course of his opium dreams: Levana and Our Ladies of Sorrow, The Daughter of Lebanon, and The Dark Interpreter. All of them archetypal figures roaming the landscape of De Quincey's unconscious.

But *Suspiria* is not merely a return journey – more lurid confessions and fantastic characters. In *Suspiria*, De Quincey begins to engage with his subject more analytically. For example, in one of the most poetic and penetrating passages he ever wrote, De Quincey considers the relationship between the brain, dreams, and 'the shadowy' (for which the word 'unconscious' might be legitimately substituted).

> The machinery for dreaming planted in the human brain was not planted for nothing. That faculty, in alliance with the mystery of darkness, is the one great tube through which man communicates with the shadowy. And the dreaming organ, in connection with the heart, the eye, and the ear, compose the magnificent apparatus which forces the infinite into the chambers of a human brain and throws dark reflections from eternities below all life upon the mirrors of the sleeping mind.

But why should De Quincey's *Confessions* and *Suspiria* be regarded as travelogues of the unconscious? In what sense did he really descend – like Orpheus – into a psychic underworld? Why make such an assertion in the first place?

Although De Quincey did not articulate a specific theory of the unconscious, his experiences (and the experiences of his romantic predecessors) do suggest a particular model of mind. A model of a mind partitioned into two distinct parts: an upper, conscious division and a lower, unconscious division (unconscious insofar as the contents of this lower division are ordinarily not available for conscious inspection). The upper mind was considered rational while the lower mind was considered irrational.

De Quincey describes descending into his sleeping mind, which conjures an image of his shrunken consciousness sinking like a diving bell into the murky depths of his own unconscious; however, such metaphors can be misleading and fail to reflect how the mechanisms responsible for producing strange experiences (such as De Quincey's) were eventually understood by nineteenth-century psychologists. The most popular explanation was nothing to do with consciousness shrinking and descending into the unconscious, but rather the contents of the unconscious expanding and rising into awareness. This was an idea with a fine philosophical pedigree. For example, Plato suggested that, in dreams, the will is unable to operate and rational control cannot be exerted over the passions. Subsequently, activities and themes emerge in the dream world that might cause considerable shame in the dreamer on waking.

If, when asleep, the conscious, rational mind is incapacitated, then any experiences (such as dreams) must reflect activity in the unconscious mind. In De Quincey's case, his consciousness, made inert by a powerful narcotic, was forced to bear mute witness to a completely unregulated eruption of unconscious material. A similar explanation could also be employed to account for bizarre hallucinations – which might be produced by the same process but then projected on to the 'screen' of the external world.

In 1861, Karl Albert Scherner published *The Life of the Dream* – further evidence of romanticism's continuing influence on works of an academic nature. Scherner believed that the language of dreams was symbolic. Moreover, he agreed that dreams were influenced by mystical or spiritual forces; however, he also recognised the influence of more mundane factors, such as physical stimulation of the sleeping body or illness. Again, like von Schubert, Scherner was an advocate of dream interpretation, but his method was somewhat prescriptive. For example, he suggested that objects such as a clarinet or knife represented the penis while narrow courtyards and staircases represented the vagina. Scherner's account of sexual symbolism is of considerable interest, as it foreshadows psychoanalytic thinking.

Exploration of the dream world continued (more in the tradition of Coleridge and De Quincey than von Schubert and Scherner) with the publication of *Dreams and the Means to Direct Them* (1867). The author was Marie-Jean Hervey de Saint-Denis, a teacher of Chinese language and literature at the Collège de France. His book contained instructions for individuals wishing to control their dreams – a skill he had acquired through systematic self-experimentation. Initially, Hervey de Saint-Denis kept detailed records of his dreams, which helped him to remember them. He then discovered that he could interrupt a dream, waking himself up in order to record unusual or particularly interesting events. As his research progressed, Hervey de Saint-Denis became increasingly self-aware during dreams. He recognised that he was dreaming and practised turning his attention towards the most interesting features of the dream environment. Finally, Hervey de Saint-Denis was able to influence the narrative of his dreams, although this degree of control was never complete. From 1896, techniques for the conscious control of dreams were refined by the Dutch psychiatrist and poet Frederick van Eaden, who coined the term *lucid dreaming* to describe the experience.

As the concept of the unconscious became consolidated, attention settled on the theoretical line dividing the mind into upper and lower chambers – the *limen* or threshold of consciousness. One of the first to consider the exact nature of this horizontal partition was the German philosopher Johann Friedrich Herbart, who discussed it at length in a two-volume work, *Knowledge Newly Founded on Experience, Metaphysics, and Mathematics* (1824–25).

In some ways, Herbart's ideas show an outmoded loyalty to the traditions of Enlightenment thinking. His model of the mind was based on principles borrowed from the physical sciences. Thus, he compared mental events (such as thoughts and perceptions) to interacting particles and sought to describe their dynamic relations using Newtonian-style mathematics; however, unlike a true representative of the Enlightenment, Herbart fully accepted the existence of the unconscious.

For Herbart, the threshold of awareness was not a smooth surface, disturbed only by the graceful ascent of memories. The limen was a plane of perpetual conflict – seething with activity. Thoughts and perceptions jostled each other, vying for a place in awareness. Stronger thoughts and impressions pushed the weaker ones below the threshold, from where they immediately fought to recover their former position. This account is peculiarly Darwinian. Animated cognitions and percepts compete with each other for a limited resource – consciousness – and only the 'fittest' survive in awareness.

Herbart can also be credited for introducing the now familiar concept of *repression*. Clearly, the 'fittest' thoughts and objects of perception must exert some kind of force to keep the weaker ones below the awareness threshold. Therefore, forgetting is not a passive process, but an active process. Forgotten information must be continuously inhibited, or it would simply fight its way back into consciousness.

The proper scientific investigation of the awareness threshold began with the advent of Gustav Fechner's *Psychophysics* – a new discipline that attempted to use mathematics to describe the limits of perception. It is ironic that Fechner, now canonised in the history of psychology as a very rigorous scientist, was also the author of works such as *Nanna, or on the Soul-life of Plants* (1848), and *Zand-Avesta, or on the Things of Heaven and the Beyond* (1851).

In 1850 Fechner started to employ mathematics to quantify perception. He determined the smallest intensity of a stimulus that could be perceived consciously. This then served as a zero point on a scale and represented what he called the *absolute threshold*. Fechner plotted the relationship between physical and subjective stimulus intensities above the absolute threshold and expressed them in a logarithmic formula, now known as Fechner's Law. This work was eventually published in 1860 as *Elements of Psychophysics*. If the unconscious were a country, then Fechner had succeeded in mapping its border; however, it was the physician and painter Karl Gustav Carus who declared that country independent in 1846, with the publication of *Psyche* – the first attempt to provide a complete and objective theory of unconscious mental life. It is the progenitor of all subsequent works that deal exclusively with the nature of the unconscious.

Instead of conceptualising the unconscious as a unitary phenomenon, Carus distinguished three different levels, each varying with respect to degree of accessibility. Moreover, he began to list what he considered to be the defining features of the unconscious. For example, he suggested that the unconscious is constantly flowing (thus, if an idea sinks into the unconscious it will continue to evolve and develop); indefatigable (unlike the conscious mind which needs periods of rest); and the unconscious has its own laws (although these are very different from those that govern conscious mental activity). Carus also showed a continuing commitment to romanticism by suggesting that the unconscious was a repository of ancient wisdom, connecting all of humanity.

A later successor to Carus was Eduard von Hartmann, whose *Philosophy of the Unconscious* (1869) overlapped with *Psyche*, but was better argued on account of frequent references to factual evidence. Like Carus, von Hartmann described different levels of the unconscious, ranging from the absolute unconscious (which corresponds with romanticism's world soul) to the psychological unconscious (which underlies the consciousness of every individual).

Although the works of Carus and von Hartmann have not been greatly influential, they mark the beginning of a new epoch. From the middle of the nineteenth century the unconscious was established as a topic worthy of independent study. A topic that merited its own literature.

In 1884, the American psychologists Charles Peirce and Joseph Jastrow published an academic paper titled 'On small differences of sensation'. It was a report on the seemingly uninspiring and dry topic of weight discrimination;

however, in fact, Peirce and Jastrow's modest study is the first investigation of what would one day be known as subliminal perception.

Peirce and Jastrow gave their experimental subjects weights and asked them to judge which was the heaviest. This task was extremely difficult because the weights were almost – but not quite – identical. Consequently, subjects were forced to make a wild guess; however, these so-called guesses proved to be remarkably accurate. Even though subjects claimed that they had no idea which was the heaviest weight, they showed a marked tendency to select the right one. This suggested that the difference between weights was being accurately determined by the unconscious mind, which was then influencing conscious guesses. Peirce and Jastrow had demonstrated that the hidden intelligence of the unconscious could be engaged in a laboratory task by feeding it information delivered below the threshold of awareness.

Back in 1702, Bernard de Fontenelle (a populariser of science) had famously anticipated 'a century that will become more enlightened day by day, so that all previous centuries will be lost in darkness by comparison'. For psychology, this was true. The workings of the mind had been illuminated for nearly a hundred years by the brilliant torch of the Enlightenment. But by 1850 the romantics had cast long shadows across the drawing rooms of Europe. The polished table clocks, which had provided the Age of Reason with such a powerful metaphor, had lost their shine. The mind was not like a well-regulated clock. It could not be dismantled and put together again by introspection. The mind was vast, deep, possibly infinite and its depths could be visited in sleep and experienced as dreams.

The unconscious had arrived.

2

Mystery and imagination

In the 1850s, the senior residents of Meersburg on the shores of Lake Constance were still able to remember the old 'aristocrat'. He had possessed, so they said, a remarkable affinity with birds. Indeed, he was often followed by a large flock – an entourage of avian disciples. Moreover, the residents recalled that the aristocrat's pet canary had become like a personal retainer, waking his master in the morning and sugaring his drink. So close was the relationship between man and bird that when the man died on 5 March 1815 the canary was heartbroken. It refused to sing or eat, and expired within a few days. The old 'aristocrat' – the birdman of Meersburg – was Franz Anton Mesmer. He died a bitter man, because *mesmerism*, his revolutionary treatment for all ailments, had been neither appreciated nor accepted by the scientific establishment. He had also discovered, inadvertently, a powerful method for gaining entry into the unconscious mind; however, unlike his many followers, Mesmer chose to ignore it.

Mesmer, the son of a game warden, was born in the Swabian town of Iznang. He studied medicine in Vienna, but demonstrated an early interest in universal forces by submitting a thesis on planetary influence and disease. After graduating, Mesmer practised medicine, but managed to escape a humdrum career (and his humble social background) by getting married in 1767 to an extremely wealthy widow, Maria Anna von Bosch – a woman descended from the German nobility. Thereafter, the young doctor's circumstances were dramatically altered. He continued to practise medicine, but did so while living the pampered life of an aristocrat. Mesmer's large estate attracted celebrity visitors such as Gluck and Haydn, and the young Wolfgang Amadeus Mozart had his first operetta, *Bastien und Bastienne*, performed in Mesmer's private garden theatre.

Between 1773 and 1774 Mesmer began to experiment with a treatment which employed magnets; however, he soon concluded that any therapeutic benefit arising from this treatment was attributable to a subtle 'fluid' or force emanating from his own body. He named this fluid 'animal magnetism'. The word 'animal', in this context, does not reflect Mesmer's assessment of his own brute power to charm but rather an etymological origin – the latin *anima* meaning soul. Although Mesmer had dispensed with the idea that magnetism per se was therapeutic, he continued to employ magnets to boost and direct his own exotic emanations.

Mesmer considered himself a product of the Enlightenment, and he was subsequently keen to develop a scientific, rational theory that could account for his clinical successes. Framing his explanation in terms of invisible fluids or forces was, in fact, entirely consistent with the most respectable science of the day. After all, no less a figure than Sir Isaac Newton had proposed a theory of universal gravitation. Moreover, Newton was known to have speculated on the existence of a universal substance (later called ether) which acted as a medium through which gravity and light might travel. Therefore, Mesmer's insistence on the existence of a weightless, invisible 'fluid', is not quite so strange when considered in a contemporary context.

Mesmer believed that physical diseases were caused by a magnetic fluid deficiency, which could be restored by provoking an attack (or seizure) called a crisis. His technique for inducing a crisis involved sitting in front of the patient (with his knees outside the patient's knees), and sweeping his hands over the patient's body (performing *grandes passes*). The patient's thumbs might be briefly held between passes. In order to flood the patient's torso with magnetic fluid, he would occasionally place one hand on the patient's abdomen and the other at the base of the patient's spine. Typically, the crisis incorporated a symptom of the underlying disease. Thus, an asthmatic might have difficulty breathing, whereas a person with epilepsy might experience convulsions (although it was the misfortune of one aristocrat, Madame de Berny, to experience a curative crisis involving an unexpected evacuation of the bowel).

Mesmer found that repeated provocation resulted in crises of decreasing intensity and a consequent reduction of symptom severity. Very occasionally, while attempting to provoke a crisis, some of Mesmer's patients looked into his eyes and became rather sleepy, an effect which was not investigated by Mesmer himself, but which nevertheless proved to be of considerable interest to some of his followers.

By 1775 Mesmer had acquired a reputation as an outstanding healer and was a firm favourite of the aristocracy, but in spite of this success he left Vienna for Paris in 1777. This sudden departure may have been due to a professional impropriety because Mesmer never saw his wife again. Ensconced in his new mansion on the Place Vendôme, Mesmer picked up where he had left off in Vienna, again becoming a fashionable doctor, but this time offering his treatment to an enthusiastic and decadent Parisian elite.

Owing to his immediate popularity, Mesmer developed a method of treating people en masse with the notorious *Baquet* – a device consisting of a large drum filled with bottles of water which Mesmer had previously magnetised. Around the drum, up to twenty patients could be arranged in order to benefit simultaneously from Mesmer's superabundance of animal magnetism. With the introduction of the *Baquet*, Mesmer's treatment procedures became increasingly like theatre: Mesmer would appear wearing a cloak decorated with alchemical symbols and then play the glass harmonica (an instrument that produced an eerie, ethereal sound). Large mirrors were erected in his magnetic salon to reflect invisible fluids and his assistants, like stage hands, were positioned to catch his convulsing patients. Contemporary woodcuts show a magus-like figure raising his hand and making women on the other side of the room swoon. It was an impressive show.

Needless to say, Mesmer soon attracted a following – mostly physicians – who eventually practised under the aegis of the Société de l'Harmonie, an unusual hybrid of college and masonic lodge dedicated to the practice of mesmerism. By 1784 Mesmer's disciples were such a conspicuous presence that Louis XVI commanded that mesmerism should be the subject of two official investigations. The first was undertaken conjointly by the Académie des Sciences and the Académie de Médecine, while the second was undertaken by the Société Royale. With respect to the latter, it is interesting to note that one of the panel members was Benjamin Franklin, not only the American ambassador, but also the inventor of Mesmer's much-loved glass harmonica.

The findings of these investigations were unequivocal: Yes, it was true that patients benefited from Mesmer's treatment, but this benefit had nothing whatsoever to do with Mesmer's animal magnetism. Treatment gains were best attributed to a psychological factor that the learned gentlemen described as 'imagination' – or what modern doctors would now call the placebo effect. A powerful expectation of improvement, once aroused in a patient, is often followed by the remission of symptoms (irrespective of the treatment's theoretical potency). For centuries, physicians have known that through the agency of the mind the body can be 'fooled' into feeling better. Mesmerism was simply a means of exploiting expectations of improvement in credulous patients.

Even though both investigations attacked Mesmer's theoretical framework, the Société de l'Harmonie continued to expand, and new branches were established all over France. However, as with any expanding empire, it was difficult to maintain sovereignty in the more remote outposts. Many of Mesmer's disciples started modifying his procedures, and some – influenced perhaps by the two enquiries – challenged his orthodoxy concerning the role of animal magnetism. Rifts opened. Factions emerged. The centre would not hold.

Mesmer could not keep the movement he had created in check. Moreover, at the same time his authority was being seriously undermined by humorists who mocked his excesses in cartoons, popular songs, and satirical plays. Mesmer's reputation was finally irreparably damaged when he was invited to attend a meeting of the Lyons Société de l'Harmonie. Prince Henry of Prussia was in Lyons on a private visit and presented himself to Mesmer as a demonstration subject. The Prince's high social rank and obvious scepticism were sufficient to overwhelm Mesmer, who discovered that his remarkable powers had chosen a most inopportune moment to desert him. He left Paris in 1785, and for many years his whereabouts were completely unknown. Even so, the Mesmeric bandwagon he had left behind rumbled on, continuing to attract the interest of a new generation of would-be mesmerists.

For the next twenty years Mesmer wandered through Europe – a man of reduced but significant means. A man who preferred the company of birds to people. When Mesmer died in 1815 most practitioners who called themselves mesmerists had no idea where he was, and, more importantly, none of their number was using his techniques to provoke therapeutic crises. They were doing something quite different.

Unlike Mesmer, Amand Marie Jacques de Chastenet, Marquis de Puységur, was a genuine aristocrat, whose ancestral home was a large castle and extensive estate near Soissons. Puységur was also a distinguished artillery officer and a keen amateur scientist (with a special interest in electricity).

He was introduced to Mesmer's doctrine of animal magnetism by his brother, Comte Antoine-Hyacinte, but from the very beginning, expressed reservations about the propriety of inducing violent crises in vulnerable patients. Indeed, he found the phenomenon undignified, if not repellent. Subsequently he experimented with a more gentle therapeutic procedure that did not necessitate a dramatic, convulsive climax.

From 1784 Puységur began offering magnetic treatments to the peasants on his estate. His first two patients were young women suffering from toothache, both of whom were cured in the absence of crises. Puységur's next patient, Victor Race (a young man with a respiratory disease), proved to be even more interesting. Victor responded to the treatment procedure in a very unusual way. After seven or eight minutes he fell into a kind of sleep, during which he was able to hold a perfectly sensible conversation, answer questions, sing songs, mimic shooting, and dance to imagined music; however, on waking, Race had no memory of any of these things.

Puységur experimented with other patients, and began to employ special instructions that encouraged 'sleep'. Still encumbered by Mesmer's theoretical framework, Puységur assumed that he had stumbled upon a new form of crisis – albeit a less dramatic form than Mesmer's convulsive original. He called his new discovery 'the perfect crisis'; however, this term was soon superseded by 'magnetic sleep' and then finally 'artificial somnambulism' (suggesting a progressive willingness to abandon Mesmer's vocabulary). Somnambulism is the medical term for sleepwalking, and Puységur had obviously recognised that the two states (magnetic sleep and the sleepwalker's trance) were close cousins. Puységur was leaning towards a psychological explanation, and eventually he concluded that artificial somnambulism had nothing to do with animal magnetism (a supposed physical force), but rather the imposition of the magnetiser's will on that of his subject.

Thus began a major rift in Mesmeric circles. Two factions emerged: traditionalists, who followed Mesmer's doctrine to the letter, and revisionists, who were more enthusiastic about Puységur's new technique and explanatory framework. The latter group abandoned the provocation of dramatic crises, focused on sending their patients to sleep, and experimented with simpler treatment methods. They also questioned the efficacy of group treatments such as the *Baquet*. Mesmerism no longer required the presence of a magus and enough props to stage an amateur production of *The Magic Flute*. In fact, mesmerism no longer required Mesmer.

In Puységur's wake, artificial somnambulism was understood to be therapeutic in several ways. Firstly, the trance state itself was thought to be beneficial because it possessed the same properties as any restorative or satisfying sleep. Secondly, when entranced, individuals were suggestible to the extent that certain symptoms could be removed by way of a simple command. However, such 'treatments' were only superficially effective, insofar as symptoms tended to reappear on waking. Finally, because artificial somnambulism was a kind of sleep, a dialogue could be established with pathogenic parts of the mind that were normally inaccessible. Thus, treatment sometimes took the form of a discussion between doctor and patient, with the patient replying to questions in his or her sleep. This presumably had a precedent in exorcism, during which priests were often called upon to bargain with evil spirits for the release of their host. Even so, the procedure merits obvious comparison with contemporary psychotherapy.

It is of some interest to note that a little-known Bavarian priest, Johann Joseph Gassner, acquired a considerable reputation as a healer by provoking therapeutic crises in his patients two years before Mesmer developed his magnetic treatment. According to Gassner, however, the therapeutic crises he provoked were caused by demonic entities, with whom he would converse before completing their exorcism with an authoritative command.

Although Puységur succeeded in transforming the practice of mesmerism, he was a reluctant revolutionary. He always considered himself a loyal follower of Mesmer and never intended to undermine the master's teachings. Indeed, he visited the great man twice, accompanied by Victor Race, to share his discoveries, but Mesmer responded coldly and was obviously unimpressed, considering artificial somnambulism to be of little significance. After all, when his own patients had 'drifted off' he had thought nothing of it.

Artificial somnambulism proved to be an extremely useful tool for probing the human mind. Indeed, Puységur's experiments revealed phenomena that could only be explained if the standard Enlightenment model – with its emphasis on rationality and transparency – was substantially revised. For example, the fact that patients could not remember what had happened to them while entranced suggested that the mind could keep secrets from itself. Clearly, Puységur's patients could not have forgotten events that had transpired only a few moments earlier. This suggested that memories of being in the trance (and associated experiences) were present in the mind, but inaccessible. Puységur went on to demonstrate the mind's capacity for self-deception even more dramatically by experimenting with what is now known as *post-hypnotic suggestion*. If an entranced individual is given a command to perform a simple behaviour (for example, 'Scratch your nose whenever you hear the word "dog"'), the command will continue to be obeyed even after waking. Such an individual will have no recollection of being given the command and will probably confabulate if asked to explain why the behaviour is being performed. Puységur had succeeded in hiding a set of instructions in the mind, thus demonstrating that there was a part of the mind which – although not available for conscious inspection – could nevertheless still influence behaviour.

A further intriguing observation of Puységur's was that some of his patients seemed to be more knowledgeable when 'asleep' than when 'awake'. For example, several individuals were able to diagnose their own problems and recommend treatments. This begged certain questions. Was artificial somnambulism allowing patients to recover information that had simply been forgotten? Or was there some vast, submerged library in the mind that could be consulted during sleep? Needless to say, the early romantic philosophers became particularly interested in Puységur's work, being inclined to believe that his patients were obtaining information from the world soul – the universal unconscious. Artificial somnambulism was quickly perceived as a possible short cut to the numinous.

Puységur also discovered that the mind was capable of not only concealing information from itself, but also concealing (or at least denying) powerful sensory experiences. Patients were told that they would not feel pain in certain areas of the body, which could then be pricked with pins and probed with heated objects without causing any discomfort. If the sensory apparatus was still functioning, then those parts of the mind allocated for the registration of pain were being shut off. A door was being closed on pain, thus keeping it outside of awareness.

Unfortunately, Puységur's investigations were interrupted by the revolution of 1789, and he spent two years in prison; however, when he was released he was able to recover his estate and go on to become the mayor of Soissons. He also continued to investigate artificial somnambulism. By the time of his death in 1835, almost all 'mesmerists' employed his procedures rather than those of Mesmer.

But there can be few individuals to whom the gods of posterity have been less generous than Puységur. In the early years of the nineteenth century 'mesmerism' (so called) continued to be endorsed by fringe medical practitioners (and a growing band of travelling entertainers); however, within a very short space of time the name of Amand-Marie-Jacques de Chastenet, Marquis de Puységur, sunk into total obscurity. Yet he had developed a method of exploring the mind which would prove to be of incalculable significance for future students of the unconscious.

From its inception to the 1840s mesmerism was never endorsed as a legitimate treatment by the medical establishment. Even Puységur's more credible methods were still regarded with considerable suspicion; however, from the 1840s mesmerism began to attract the attention of several British doctors, whose scientific credentials granted it a degree of vicarious respectability.

John Elliotson was appointed professor of medicine at the University of London in 1831. He founded University College Hospital, established a link between pollen and hay fever, and pioneered iodine treatment for goitre. In addition, he was the first doctor in England to make extensive use of the stethoscope, an instrument that many of his colleagues were happy to dismiss as a European fad.

Elliotson became interested in animal magnetism after attending various staged demonstrations conducted by visiting continental mesmerists. He was particularly impressed by the induction of anaesthesia. When in a trance state, subjects could be pinched or have their nostrils packed with snuff without showing any signs of discomfort. Such phenomena suggested that mesmerism could be used to moderate pain during surgery (the very first recorded use of ether was not until 1842 and that of nitrous oxide, 1844). In 1843, Elliotson published a pamphlet titled 'Numerous Cases of Surgical Operations without Pain in the Mesmeric State'. This represented the first attempt to collect together existing documented cases of mesmeric surgical anaesthesia, the first of which was a mastectomy performed by Jules Cloquet in 1829.

Unfortunately, Elliotson's reputation was diminished after his involvement with the Okey sisters – two adolescent patients who proved to be remarkably compliant experimental subjects. Elliotson used them to demonstrate the power of 'animal magnetism', first on the wards and then in the public theatre of the hospital. The girls could be made to go rigid, swoon, and perform astonishing feats of strength. Unfortunately, Elliotson's demonstrations degenerated into an undignified stage show, attracting large audiences which included not only doctors, but also aristocrats, members of parliament, writers, and anybody of sufficient social rank to gain entry. The behaviour of the Okey sisters became increasingly idiosyncratic and unpredictable. They entertained Elliotson's audiences by adopting a peculiar, childish mode of speech, by using bad language, and showing little or no respect to even the most distinguished members of the gathering. Elliotson's demonstrations became a combination of slapstick comedy and mild sexual titillation. A subsequent investigation concluded that Elliotson was being duped by two crafty imposters – an allegation that Elliotson simply refused to accept. Eventually the hospital authorities informed Elliotson that he should refrain from practising mesmerism in the hospital and that he should discharge the elder Okey sister immediately. Elliotson responded by resigning his post.

Among the great and the good attending Elliotson's demonstrations had been Charles Dickens. The two men became firm friends, and Elliotson taught Dickens how to mesmerise. The author subsequently experimented with his own family, and later successfully healed some of his associates. It has been suggested that Dickens' extraordinary power to captivate large audiences at public readings of his work was in part due to the exercise of his mesmeric gift. Dickens insisted that his audience should always be able to see his face, and if the audience wasn't responsive he was quick to complain that they were not 'magnetic'.

Although Elliotson was spurned by his university colleagues, he was warmly accepted in Dickens' elevated social circle and his medical practice prospered. In spite of his unwise association with the Okey sisters, Professor Elliotson's endorsement of mesmerism was a turning point. Mesmerism had embarked on the road to respectability.

After the publication of Elliotson's pamphlet on the use of mesmerism as a surgical anaesthetic, news began to reach London of a lone enthusiast practising in Bengal. This was James Esdaile, a Scottish surgeon with the East India Company, who ran the so called Native Hospital. Using a mesmeric trance state to induce anaesthesia, Esdaile had conducted several major operations with good results. These included arm and breast amputations as well as the removal of numerous scrotal tumours (which were endemic in the Bengali population). One of these tumours weighed more than the patient, and had to be manipulated with a rope-and-pulley system attached to the rafters. Clearly, Esdaile was a man with considerable nerve.

Remarkably, Esdaile had never seen a professional mesmerist induce a trance state. He based his own procedures on an account given to him by a friend, and subsequently developed a somewhat idiosyncratic technique that accommodated local influences (such as yogic breathing and stroking). Although his technique was improvised, an official investigation was impressed by his results and in due course he became the founder and superintendent of The Calcutta Mesmeric Hospital. He left India in 1851 and returned to Scotland, where he lived a relatively uneventful life of semi-retirement.

During his six years in India Esdaile performed several thousand operations on mesmerised patients. Moreover, he kept careful records and tabulated his successful results. Only sixteen deaths were reported, at a time when only 50 per cent of surgical patients were expected to survive. Sadly, Esdaile's findings were not given the consideration and exposure they deserved because of the British medical establishment's racist views. It was suggested that 'natives' of the subcontinent were so different from Europeans that they might actually enjoy surgical procedures. Therefore, it was impossible to assess the effectiveness of mesmeric anaesthesia on the basis of Esdaile's work.

It was yet another Scottish surgeon, James Braid, who finally earned mesmerism scientific respectability; however, the task was not an easy one, and to achieve it the concept of mesmerism had to be thoroughly rehabilitated. Even a name change was necessary.

Like Elliotson, Braid's first experience of mesmerism was at a public demonstration – on 13 November 1841. Braid was so intrigued by what he saw that, one week later, he returned to see the whole thing again (when it was repeated by popular demand). Braid was convinced that he had witnessed a genuine phenomenon; however, he was not satisfied with any of the existing explanations. Even some sixty years after Mesmer's heyday there was still talk of emanations and magnetic fluids (particularly among stage performers who sought to sensationalise their act).

At both demonstrations Braid had observed that during 'nervous sleep' the subject's eyes remained firmly closed. He subsequently concluded that the trance state had been induced by neuromuscular exhaustion, brought about through protracted staring. Two days later, to test his theory, Braid invited a dinner-party guest to stare, without blinking, at the top of a wine bottle. Within minutes the man was asleep. Braid subsequently repeated this experiment with his wife and manservant, who also obligingly fell asleep.

Braid's initial approach, then, was to understand nervous sleep primarily as a physiological phenomenon. He later elaborated his account to the extent that he acknowledged the importance of a psychological factor – 'focus of attention' – as another necessary precipitant of nervous sleep. Braid's explanatory framework is very straightforward and marks a radical departure from all that went before. Concepts such as animal magnetism, or Puységur's 'imposition of will', are completely rejected in favour of more basic elements. Braid subsequently spent the next eighteen years of his life researching nervous sleep, and in 1843 renamed it *neurypnology*. He later chose another name, hypnosis, which proved to be so 'catchy' it soon replaced mesmerism, artificial somnambulism, and nervous sleep – going on to achieve international currency.

Braid used hypnosis to treat a wide range of problems, from spinal curvature to epilepsy. Moreover, he often provided a rationale for his successes that demonstrated his fealty to respectable science. Thus, he claimed to be able to cure deafness because the auditory nerve – an object familiar to neurologists – could be excited under hypnosis. Even so, Braid's successes probably owed as much to the placebo effect as did Mesmer's. Braid's explanations were simply much more attractive to the conservative medical establishment and he was subsequently able to publish his findings in mainstream academic journals. As a direct result of Braid's publications, hypnosis was rescued from the world of quacks, mountebanks, and music hall to be delivered safely into the hands of neurologists and physiologists; but just as the scientists were becoming used to the idea that hypnosis was respectable after all, it was more or less hijacked by a burgeoning fringe religion – spiritualism.

Among spiritualists, communication with the dead was usually accomplished after entering a trance state; but such trance states could also be conceptualised as 'self-hypnosis'. Moreover, many hypnotic subjects began spontaneously to report receiving messages from the spirit world. Be that as it may, even though the culture of spiritualism was steeped in superstition and absurdities, certain previously unseen 'psychological' phenomena emerged that captured the attention of the academic and medical communities. The unexpected result was even greater scientific interest in the function and capabilities of the unconscious mind.

The spiritualist movement was inspired by the life and works of the Swedish mystic Emanuel Swedenborg. He wrote a number of exegetical works – allegedly under the tutelage of spirits and angels – and described transcendental journeys in books such as *On Heaven and its Wonders and on Hell* (1758). After his death his visionary teachings were promulgated by a religious sect – the Swedenborgians. Congregations were soon established in northern Europe and, shortly after, America. By the mid nineteenth century the concept of communicating with the dead had become popularised under the banner of spiritualism, whose priest-class – mediums – typically entered a trance state to receive information from favoured spirit guides.

As the movement grew, spiritualist meetings became increasingly theatrical, seances being often enlivened by table-turning, rapping, and the occasional levitation. But such occurrences could easily be dismissed as conjuring tricks (stage magic had become something of an art during the course of the eighteenth century); however, other spiritualist phenomena were far more difficult to explain away – most notably, *automatic writing* and *automatic drawing*.

In the 1840s a number of spiritualists began to produce literary and artistic works that had been completed while entranced. It was subsequently claimed that these works were accomplished under the influence of spirit guides. Because the writing and drawing was performed without volition, the term 'automatic' was employed to describe the manner in which they were executed. A perplexing feature of these automatic phenomena was that sometimes works of outstanding quality were produced by individuals who had received little or no formal education. For example, Andrew Jackson Davis, the son of a New York leather worker, wrote numerous books of scientific and philosophical interest, including *Principles of Nature* (1847), which became a best-seller. Moreover, examples of automatic drawing were so technically proficient – and distinctive – the 'automatic style' exerted an influence on the early symbolists.

For those who wished to account for automatic phenomena without recourse to spirit communication, the unconscious became an invaluable explanatory concept. In 1854, for example, Michel Chevreul suggested that all messages from the spirit world might be nothing more than a transliteration of unconscious thought. Chevreul – a sceptic with impeccable credentials – had already demonstrated in 1833 that the movements of the divining rod were unconsciously directed by the dowser.

Automatic writing and drawing were subsequently considered as further examples of the richness of unconscious life. As the romantics had suggested, the engine of imagination was probably submerged below the awareness threshold. Therefore, entranced mediums were simply surrendering control of their hands to the creative genius of the unconscious; however, the fact that many mediums claimed to be taking dictation or receiving instruction from spirit guides suggested another intriguing possibility – that parts of the unconscious could evolve into fully fledged secondary (or even tertiary) personalities. The unconscious might actually be inhabited by lesser selves – amalgams of inaccessible memories that had become organised around a kind of proto-identity.

This arresting idea (that parts of the unconscious could acquire the properties of an independent identity) resonated with a rare medical phenomenon that had been observed (but never explained) since the end of the eighteenth century – split or *multiple personality*.

As early as 1791, Eberhardt Gmelin reported a case of what he described as 'exchanged personality'. The case in question was a young German woman who regularly swapped her Teutonic sensibilities for those of a French aristocrat. When her alter ego took over she spoke only French and adopted gallic affectations, and when she reverted she had no knowledge of events witnessed as a Frenchwoman. It should be noted that, even in this pioneering case study, there is no suggestion that the woman was the victim of possession. Multiple personality was understood to be a purely psychological phenomenon.

After Gmelin a few cases of multiple personality were reported in the literature, but the proper scientific study of the phenomenon did not begin until the publication in 1840 of a monograph by a general practitioner – Antoine Despine. In this work, Despine reported the case of an eleven-year-old girl suffering from paralysis. During the course of her treatment she developed the habit of slipping into an altered state of consciousness, in which she became ill-mannered but also miraculously recovered the use of her legs. Eventually, her polite and impolite personalities fused together. Despine's study failed to arouse a great deal of interest in academic circles, but from the 1840s onwards cases of multiple personality were recognised and reported with increasing frequency.

As the nineteenth century progressed, enough cases were reported to discriminate between different manifestations of multiple personality. For example, in some cases two personalities each knew about the other whereas in other cases both personalities were mutually amnesic (i.e. ignorant of each other's existence). The relationship between personalities could also be asymmetric – with only one being amnesic of the other. Finally, a number of cases were known to have several sub-personalities: a community of identities in which each sub-personality might have complete, partial, or no knowledge of every other sub-personality; thus, personality A might be fully aware of B, partially aware of C, and completely ignorant of D. Whereas personality D might be fully aware of A, partially aware of B, and completely ignorant of C. And so on.

These findings had enormous implications for the evolving model of the human mind. Whereas initially one horizontal division had been proposed, to separate upper and lower regions (conscious and unconscious), it now seemed that a multiplicity of divisions was possible. An almost infinite number of complete and partial partitions could be erected, enabling the mind to accomplish endless permutations of self-deception. In the latter half of the nineteenth century two distinct models of the mind became consolidated under the banners of *dipsychism* and *polypsychism*.

Advocates of dipsychism believed in the double ego – the presence of a secondary personality, largely concealed in the unconscious. The most contentious issue surrounding dipsychism concerned whether the hidden mind received information exclusively through the gateway of consciousness or whether information could arrive by other pathways. In the former model, the hidden mind was organised around forgotten information and trace perceptual experiences; but in the latter, it was suggested that the hidden mind might develop from more exotic material – the most likely source being mystical in nature (such as the romantic world soul).

Polypsychism was an altogether more complicated idea. Advocates of this approach viewed the human psyche as a community of lesser minds, whose operation was co-ordinated by a master (or executive) mind. The arrangement might be compared to a classical orchestra. Each of the individual sections – for example, strings, wind, or brass – can function independently; however, they are usually united under the conductor's baton. In polypsychism, lesser minds can function independently like the sections of an orchestra. They possess a specialist repertoire (unique memories and unconscious regions); however, these lesser minds usually work together under the watchful eye of the master mind. This overseeing mind – the conductor – is the identity we recognise as ourselves when we introspect. Obviously, polypsychism provided the best account of complex multiple-personality cases. Extending the orchestral analogy, the conductor might be temporarily indisposed, allowing the first violin to leap on to the podium and turn an orchestral concert into a string concert. Needless to say, the heads of the other sections might also be capable of hijacking the programme in much the same way.

Eventually scientific investigators concluded that spiritualist phenomena could be explained entirely within the frameworks offered by either dipsychism or polypsychism. For example, the physician Théodore Flournoy undertook a five-year study of the medium Helen Smith. His results were published in *From India to the Planet Mars: A Study of a Case of Somnambulism with Glossolalia* (1900). Flournoy concluded that Helen Smith's revelations were merely 'romances of the subliminal imagination', derived largely from forgotten sources (for example, books read as a child). He subsequently coined the term *cryptomnesia* to describe the phenomenon. Flournoy also concluded that Helen Smith's spirit guide, Leopold, was merely an unconscious sub-personality.

Romanticism had established an intellectual climate which favoured the recognition of unconscious mental activity. Subsequently, hypnotism, phenomena associated with spiritualism, and reports of multiple personality, reinforced the view that any model of mind that failed to acknowledge the unconscious must be incomplete. Indeed, the concept of unconscious mental activity had become an essential explanatory vehicle – at least for those who professed a scientific outlook; however, new ideas about the mind did not respect the boundary between art and science, and throughout the nineteenth century many literary works appeared which were distinctly 'psychological'.

Mesmerism inspired an entire genre. In addition to numerous lesser writers, giants such as Balzac, Dumas, Poe, Browning, and de Maupassant all wrote works featuring mesmerism. This tradition culminated with the publication of George du Maurier's *Trilby* (1894), a novel in which the eponymous heroine enjoys a brilliant (but ultimately doomed) singing career, under the hypnotic influence of her wicked mentor Svengali.

Yet more interesting are those works that seem to be inspired by the concept of split personality. In James Hogg's *The Private Memoirs and Confessions of a Justified Sinner* (1824), the anti-hero Robert Wringhim meets a sinister stranger called Gil-Martin, who represents Satan, but may equally represent Wringhim's own dark side. A few decades later, Hans Christian Andersen wrote an enigmatic fairy story called *The Shadow* (1847), in which a scholar becomes separated from his shadow. When the shadow returns, they exchange roles, which proves fatal for the scholar. A similar theme emerges in Fyodor Dostoevsky's early novel *The Double* (1845): Mr Golyadkin, a government clerk, encounters a pernicious doppelgänger whose actual existence is never properly established and who may be the embodiment of everything the 'real' Mr Golyadkin hates (and fears) about himself.

The definitive novel of this genre is, of course, Robert Louis Stevenson's much celebrated *The Strange Case of Dr Jekyll and Mr Hyde* (1886). By using drugs, Jekyll liberates his dark side, which surfaces as the repellent Mr Hyde in whose person Jekyll commits acts of brutality and murder. Yet even this masterpiece was not the final word. Very late in the nineteenth century the theme of dipsychism was still being explored in literature. As, for example, in Oscar Wilde's *The Picture of Dorian Gray* (1891) – the tale of a beautiful young man who does not age while his portrait is ravaged by time and the consequences of a decadent lifestyle.

Clearly, the 'psychological' novels of the nineteenth century preserve the romantic tradition: the human mind is fathomless and the unconscious a secret chamber. Unlocking the unconscious can release elemental forces, some of which can assume identities. Thus, we are provided with a new demonology – sanctioned by science. And the new demons of the unconscious, like their hellish counterparts, are just as capable of taking control. The unconscious is the source of everything imaginable, but, in a world where anything can be imagined, to open the sluice gates of the unconscious is always potentially dangerous. By the late nineteenth century the unconscious had become like Pandora's box – something fascinating but something to be handled with care. Something that merited a plethora of cautionary tales.

This rather disconcerting view of the unconscious also emerges in the philosophical works of Friedrich Nietzsche, whose writings appeared at roughly the same time as several of the literary masterpieces cited above. Friedrich Nietzsche (a man whose public image has suffered unfairly because of friendship with Wagner and the indiscriminate approval of Nazi 'intellectuals'), viewed man as a self-deceiving creature with compromised insight, largely ruled by a chaotic, primitive unconscious – a maelstrom of confused thoughts, feelings, and instincts.

Throughout the nineteenth century many 'scientific' commentaries on the unconscious were still steeped in the romantic tradition; however, at the same time, another approach to the unconscious was also beginning to develop. A cooler approach. Less grandiose and less 'literary'. By contemporary standards, more recognisably scientific.

In 1786 the physician and physicist Luigi Galvani produced muscular contractions in a frog by prodding its nerves with a pair of scissors during an electrical storm. This discovery initiated a programme of research which eventually resulted in the publication of *Commentary on the Effect of Electricity on Muscular Motion* (1791). In this work, Galvani proposed that muscle contraction was produced by an electrical 'fluid' that originated in the brain and flowed through the nervous system.

From earliest times animation has always been associated with spirit forces and the soul. In *De Anima*, Aristotle discusses how the soul enables animals to move, by releasing *pneuma* (or spirit) which runs through the body. By getting a frog's leg to twitch with metal probes and pieces of wire, Galvani was seizing control of a process that had hitherto been strongly associated with the divine. It was God who permitted animals to move, by equipping them not only with the means, but also the will to do so.

It was assumed that 'animals' – from insects to domestic pets – exercised a form of choice; early descriptions of animal behaviour are full of charming, anthropomorphic observations that reflect this view. For example, the diarist John Evelyn was moved to write the following on Italian hunting spiders:

> I have beheld them instructing their young ones how to hunt, which they would sometimes discipline for not well observing. But when any of the old ones did (as sometimes) miss a leap, they would run out of the field, and hide them in their crannies, as ashamed, and haply not be seen abroad for four or five hours after.

Galvani had produced movement without volition. A physical demonstration of 'unconscious' action.

The word 'reflex' was introduced into biology by the neurologist Marshall Hall and was used to describe the 'automatic' or involuntary response of a muscle, or group of muscles, to a given stimulus; however, a number of physiologists and neurologists began to consider the possibility that certain mental phenomena might be analogous to automatic reflexes. For example, involuntary memory (as exemplified by the common and frustrating experiences of being unable to remember a fact or name which at a later date effortlessly pops into awareness). Thomas Laycock (a professor of medicine at Edinburgh) suggested that such phenomena reflected the 'reflex function of the brain', while a contemporary rival, the physiologist Benjamin Carpenter, referred to these phenomena as 'unconscious cerebration'.

In 1876, Carpenter observed that volition seems to play a very minor role in the execution of even the most simple behaviours – which are largely automatic. For example, numerous muscles must be activated and organised for an individual to produce a single musical note or syllable. Achieving the exact configuration is a complex task. Yet this is routinely accomplished without any mental effort:

> We simply conceive the tone or syllable we wish to utter and say to our automatic self, do this and the well trained automaton executes it. What we

will is not to throw this or that muscle into action, but to produce a certain preconceived result.

A few years earlier, in 1874, the great Victorian populariser of science Thomas Henry Huxley also considered the modest role of volition in initiating actions, although he reached an altogether more dramatic conclusion. In his article 'On the hypothesis that animals are automata', he argued that all animals – including man – depend largely on automatic processes to function. Somewhat controversially, he insisted:

> The feeling we call volition is not the cause of the voluntary act, but simply the symbol in consciousness of the stage of the brain which is the immediate cause of the act. Like the steam whistle which signals but doesn't cause the starting of the locomotive.

So, consciousness was an epiphenomenon – insubstantial, illusory. Meanwhile, down below, in the mind's engine room, a machinery comprised of automatic associations and neural reflexes was generating cognition, emotion, and complex actions. With Carpenter and Huxley we begin to see the emergence of an extraordinary post-industrial unconscious, populated not by exotic spirit guides, poetic apparitions, and evil sub-personalities – but by robots.

For those who subscribed to the view that most of human behaviour was produced by automatic processes, the obedience of hypnotic subjects was now readily understood. With the power of volition anaesthetised, the will of the hypnotist became a kind of surrogate will. The human body was like an empty carriage, with a new occupant in the driver's seat.

The concept of automatic and unconscious processes in the brain was also employed to explain certain features of visual perception. For example, how is it that we know when objects are moving away from us? The physiologist Hermann von Helmholtz suggested that initially, to make sense of the world, human beings must learn rules. For example, 'The size of an object's image varies inversely with its distance from the eye'. Subsequently, if an individual perceives a ball getting smaller, then it follows that the ball is receding; however, most people are not aware of engaging in laborious syllogisms to work out the direction of a ball during a game of tennis. Helmholtz suggested that the basic rules of perception are so well learned they operate automatically. Thus, we understand the visual environment with the help of a process Helmholtz called *unconscious inference*.

Clearly, the unconscious, as understood by the likes of Carpenter, Huxley, and Helmholtz, was a rather different unconscious from that as understood by De Quincey, Carus, or Flournoy. The romantic conception of the unconscious was many things, but principally it was an independent agency – within the mind. It could create great works of art, conjure visions in the sensory apparatus, or organise itself around an identity. This other, cooler version of the unconscious was much more 'mechanistic'. It was almost as though the Enlightenment table clock had made a comeback. But this time, the mind didn't only operate *like* a clock, the mind was actually *generated* by a clock. Beneath the awareness threshold, impersonal and disinterested processes – processes that worked like clockwork or neural reflexes – were responsible for recollection, perception, and action. This new group of theorists seemed to be suggesting that Pandora's box contained no intelligence, no wise agencies – merely brain circuits that twitched insensibly, like frog's legs prodded with wire. Could they possibly be right? It would be almost a hundred years before this kind of thinking was given the serious consideration it properly deserved.

After the mid nineteenth century and the sanitization of hypnotism, its use as both a treatment and an exploratory tool became more frequent in medical circles – particularly among neurologists and psychiatrists. The phenomenon of post-hypnotic suggestion had become well established. It had been repeatedly demonstrated that instructions capable of influencing behaviour could be planted in the unconscious. This led some practitioners to consider the role of latent ideas in the formation of behavioural symptoms. Could, for example, certain ideas enter the mind and work their way down into the unconscious? Could the mind be poisoned – as it were – from below the awareness threshold, and could such a mechanism explain mental illness?

In 1873 the physician, Auguste Ambroise Liebeault, wrote that an idea planted in the mind of a hypnotised individual might become lodged inside – or 'fixed' in the terminology of the day.

> … while the mind is occupied with the daily actions of normal life which the subject accomplishes consciously and of his own free will, some of the ideas suggested in that former passive state continue their hidden movement. No obstacle can hinder them in their fatal course …

If mental illness was caused by pernicious ideas that had become fixed in the unconscious, then how could they be discovered? How could they be brought into awareness? And how could they be removed? It was in pursuit of the answers to questions such as these that psychotherapy was born.

In the early 1880s psychotherapy did not exist (and hypnosis had proved to be an unreliable treatment). By the end of the 1880s, however, everything would change. The unconscious was to have a central role to play in *fin-de-siècle* explanations of mental illness. Moreover, psychotherapy would be the institution and vehicle through which the unconscious would become the most celebrated of all psychological concepts.

3

The philosophy teacher

In 1885 something very strange was happening in Le Havre. That autumn a young philosophy teacher, Pierre Janet, had begun a remarkable set of experiments on a woman called (in order to preserve her anonymity) Madame B. His findings were presented before the scientific community in Paris on 30 November, when his paper was read at a meeting of the Société de Psychologie Physiologique. The eminent neurologist and director of the Salpêtrière hospital, Jean Martin Charcot, was in the chair. For reasons which are now unknown, the young man himself did not attend; instead, his uncle Paul (also a philosopher) delivered the paper on his nephew's behalf. It was unclear how such a distinguished audience would receive this unusual presentation; however, when Paul Janet closed his address, even the most sceptical members of the Société were impressed, and perplexed, by the young man's findings. On reflection, it was decided that the results of his investigation should not be made available to the general public; however, even the most discreet members of the Société found it difficult to maintain a dignified silence. They talked among themselves, then to their colleagues, and finally to their friends. In fact, they talked so much about the philosophy teacher's findings that, in April of the following year, a delegation representing the newly formed Society for Psychical Research set off from England for Le Havre for the purpose of finding out what, exactly, had been going on.

The delegation was led by Frederick Myers, a fellow of Trinity College Cambridge and one of the founders of the Society for Psychical Research, an organisation committed to the scientific investigation of paranormal (or psychical) phenomena. Myers was by no means a gullible man, and he had read Janet's 1885 publication 'Note sur quelques phénomènes de somnambulisme' with due detachment. In this article, written for the *Bulletin de la Société*, Janet reported his controversial findings in full. They appeared to provide the reader with evidence for 'suggestion at a distance', that is the induction of a hypnotic trance state using telepathy. Myers was aware that many early hypnotists claimed that their powers could extend over great distances. Puységur, for example, conducted a demonstration in which – without saying a single word – he directed one of his patients to a chair and then instructed her to bring him certain objects. He had apparently controlled her behaviour by exercising his will alone; however, such phenomena had never been the subject of a thorough scientific investigation. Janet's work, confidently presented before the Paris intelligentsia, represented a bold assertion that this hitherto dubious phenomenon was both real and demonstrable. Initial reluctance to publicise Janet's work was due to fear of attracting the unwelcome interest of overzealous enthusiasts. At that time, yet another craze for stage hypnotism and sensational phenomena was sweeping France and the Le Havre area was particularly affected. Needless to say, the delegation from England were not considered to be dilettantes or mere followers of fashion.

When Myers arrived in Le Havre he was introduced to the young philosophy teacher, a man with daunting academic credentials. Janet was a graduate of the École Normale Supérieure, a school for the intellectual elite of France. 'Normalians' were groomed for professorship, and although exempt from military service were pledged to teach for ten years. Janet had taken up his appointment at Le Havre in 1883; however, although committed to the teaching of philosophy, he had long since nurtured an interest in medicine and psychiatry. Over the years this had become something of a passion. Whenever he visited his family in Paris he invariably spent a significant proportion of his time seeing patients with his brother Jules (then, a medical student). On returning to Le Havre, he chose to spend most of his leisure time as a voluntary worker at the local hospital. Janet was eager to start work on his *doctorat ès lettres* thesis, and it was almost inevitable that he would want to research an area relevant to medicine and psychiatry. He had been fortunate enough to have made the acquaintance of Dr Gibert, a local physician who had, among his many other patients, a woman who could be hypnotised by an act of will. Janet's interest was aroused immediately.

Madame B lived with Dr Gibert's sister at a house called Pavillon. She was described by Myers as a simple and honest character, a 'heavy, middle aged, peasant woman, with a patient, stolid expression, and a very limited intelligence and vocabulary'. Dr Gibert had frequently experimented with hypnotism as a method of treatment, and had, by chance, discovered her unique susceptibility to the trance state; however, he had not undertaken any rigorous and systematic investigations of 'suggestion at a distance' and was happy to leave Madame B in the hands of his capable and enthusiastic junior for this purpose.

Myers was eager to see Janet's original notes on the case. The young man obliged without question. Myers was struck by his scrupulosity. Each observation had been noted in a meticulous, careful hand. This was consistent with the man's general bearing. He was quite short in stature, lean, with dark eyes and a trimmed, pointed beard. Fastidious, sharp. Very much a Parisian.

Janet's notes made extremely interesting reading. He had induced a trance state in Madame B by holding her hand and applying gentle pressure. Even a slight pressure from his thumb would suffice; however, he noticed that unless he was 'willing' Madame B to sleep, nothing happened. Touch without intention was ineffective. Further proof of this phenomenon was obtained when Janet sat with Madame B while Dr Gibert 'willed' her to sleep from an adjoining room. Janet's touch, without intention, did nothing. It was only when Gibert began to will Madame B to sleep that the desired outcome ensued.

Myers found an even more compelling entry dated 3 October 1885. Gibert had attempted to send Madame B to sleep from a distance of half a mile; however, when Janet went to investigate he found Madame B awake. She was somewhat indignant: 'I know very well that M. Gibert tried to put me to sleep, but when I felt him I looked for some water, and put my hands in cold water. I don't want people to put me to sleep in that way; it puts me out, and makes me look silly.' Further entries confirmed that Madame B knew whether it was Gibert or Janet who was attempting to make her sleep. This seemed to support an idea originally suggested by the early pioneers of hypnotism, that a strange bond called the *rapport* linked the hypnotist with his subject.

On 14 October 1885 Gibert had induced a trance in Madame B from a distance of two-thirds of a mile, at an hour suggested by a third person but not disclosed to Janet. When Janet recorded the time that Madame B fell asleep it coincided with the appointed hour. Myers recognised that these findings excluded the possibility that Madame B was detecting subtle and unintentional signs of anticipation from Janet. Given that he was ignorant of Gibert's intentions, this was clearly an impossibility.

Although impressed by Janet's notes, Myers was not entirely convinced. He had come to witness the phenomenon, not read about it. Therefore, it was agreed that he should participate in a series of experiments conducted between 20 and 24 April 1886. The first set of experiments were inconclusive; however, after 22 April Myers became increasingly certain that the phenomenon was genuine.

On the evening of 22 April Dr Gibert pressed his forehead against Madame B's and planted a 'mental suggestion' in her mind. This mental suggestion had been chosen by Myers. As Gibert undertook his task, Myers made sure that Gibert was silent, and made no gestures that might be interpreted as disguised directions. Myers had suggested to Gibert that he instruct Madame B to examine a photographic album in the salon of Pavillon at precisely eleven o'clock the following day. Gibert and Janet were very familiar with Madame

B's daily round of activities. Most mornings she habitually sat sewing in the kitchen or in her bedroom; looking at photographs would be an unusual departure from her rather monotonous and predictable routine.

On the morning of the 23rd, Janet took Myers and his party to Pavillon. They waited in the room opposite the salon and at eleven o'clock, Madame B entered the salon and wandered about with 'an anxious, preoccupied air'. The men entered the salon after ten minutes and found Madame B in a hyp- notic state. Her eyes were open, but they were fixed. They left again, and watched her through the partially opened door. At twenty minutes past eleven she began to handle some photographic albums. She sat on the sofa and stared at the photographs until she drifted off into a deep sleep.

Later that day Janet attempted to make Madame B sleep at four-thirty (from his own house). At about five o'clock Myers and his party entered the salon and found her, eyes shut, asleep but sewing vigorously. She began to mumble, addressing Janet by name. As Myers listened carefully, he heard her say: 'It is you who made me sleep at half past four.' Those present tried to convince her that it was Dr Gibert who had made her sleep, but the woman maintained that it was Janet.

After many demonstrations of this nature Myers concluded that the phe- nomenon under investigation was genuine. He believed that he was able to exclude the possibility of fraud, accidental coincidence, and the use of coded directions or gestures. Moreover, he was persuaded that none of the Le Havre group were eager to exploit their 'celebrity'. Quite the contrary, in fact. Indeed, Madame B herself seemed to be finding the whole procedure irritat- ing, unwelcome, and demeaning. When Myers returned to England he gave an account of his visit to Le Havre in a paper titled: 'On telepathic hypnotism and its relation to other forms of hypnotic suggestion'. His visit to France had been most productive.

Myers later wrote a work titled *Subliminal Consciousness* (1892), in which he accounted for a range of phenomena – both normal and abnormal – by impli- cating a division of mind operating beyond awareness. He was convinced that hypnosis was a particularly useful tool for the study of subliminal conscious- ness, and urged psychologists to study the hidden mind within an experimen- tal framework. Although he is now remembered primarily as a founding father of psychical research, his more general writings on the mind and its relation to matter and human personality are of considerable historical importance.

But what of the philosophy teacher? What was his response to widespread interest and acclaim? To his enormous chagrin, he found that he was being inaccurately quoted. His experiments were being described by individuals who had simply not read his work properly. Moreover, unlike Myers, he was far from certain that his studies *did* constitute evidence for telepathic hypnotism. Ironically, he wasn't convinced. Showing quite remarkable restraint for an individual who might have, at that point, quite easily commanded massive public and academic interest, he simply abandoned further research into telepathy. Instead, he restricted himself to the more fundamental phenomena of hypnotism and suggestion. For the rest of his life Janet remained distrustful and sceptical of any claims for the existence of paranormal phenomena. His work with Madame B is now almost totally forgotten.

Although Janet did not realise it at the time, he was setting a precedent. He was exhibiting a 'gift' for obscurity. Like some curious reverse social alchemist, he had an uncanny knack of turning the gold of fame into the base metal of anonymity. It can be argued, without hyperbole, that Janet's numerous books and publications easily secure him a place in the pantheon of great thinkers about the mind. It might even be argued that his works on psychology and psychiatry rank with (and perhaps even surpass in importance) the acknowledged greatest. Unfortunately, his gift for obscurity has rendered his name virtually meaningless outside his native France. Few have heard of him. Psychology has forgotten its Newton.

Janet was born on 30 May 1859 at 46 rue Madame, Paris. His father, Jules, was a legal editor, and his mother, Fanny, the daughter of a Strasbourg building contractor. The marriage produced three children: Pierre, Jules, and Margaret. Young Pierre was reasonably intelligent, but it was not until his adolescence that his true potential was recognised under somewhat curious circumstances. At the age of fifteen Janet became severely depressed and his education was interrupted for several months. When his depression lifted he threw himself into his work, becoming an outstanding student. After passing his baccalaureate examination he succeeded in securing a place at the École Normale Superiéure.

By the age of twenty-three Janet seemed destined to distinguish himself as a philosopher; however, the intellectual climate of the time encouraged the consideration of many other possibilities. Advances in science and technology promised to change the world beyond recognition (indeed, Paris was less than five years away from having its skyline transformed by Eiffel's engineering marvel that would become one of its defining features). The most controversial advances, however, were occurring in the field of psychology.

In 1882, while Janet was at the École Normale, the eminent neurologist Charcot delivered a highly influential lecture at the Académie des Sciences. He dismissed reservations concerning the credibility of hypnosis and insisted that it could be used to understand mental illnesses, most notably, *hysteria*. In the 1880s hysteria was a relatively common diagnosis. Patients would experience pains or numbness (anaesthesiae) in different parts of the body, adopt unusual postures, and become emotionally disturbed for no apparent reason. These symptoms tended to occur during acute episodes (also called attacks or crises) although minor symptoms such as muscular contractions or numbness could be a permanent feature. Between episodes, patients could make a full recovery; however, their recollection of events that occurred during attacks was usually very poor. Physical investigations failed to reveal any biological problems or diseases of the nervous system. Hysteria was a mystery. When Charcot claimed to have found some answers he became the talk of Paris. The young Janet, like many of his contemporaries at the École Normale, began to formulate a long-term plan to study medicine.

Charcot became what the French described as a *prince de la science*. He was a distinguished investigator, whose wealth and reputation gave him influence that extended well beyond the parameters of academic medicine. He once arranged an informal meeting that resulted in the formation of the Franco-Russian alliance; however, he was most famous for his dramatic lectures which (like Elliotson's antics with the Okey sisters) became major cultural events, attracting visiting foreign dignitaries, writers, celebrated actresses, and, needless to say, members of the medical and scientific community.

Charcot would often use patients to illustrate key points, beguiling his audience by showing how hysterical symptoms could be created and removed by the power of suggestion. In spite of his enormous historical importance, it is difficult to assess the value of Charcot's work. Indeed, it has been suggested that, at least to some degree, Charcot, his students, and his patients, all colluded with each other to keep the 'circus' going. The show that they devised was very successful, and ensured celebrity for all who took part. One patient, for example, became known as the 'Queen of the Hysterics'. Charcot believed that much could be learned from the study of faith healing and perhaps, in the final reckoning, Charcot's greatest clinical tool was not hypnotism but showmanship. Like Mesmer before him, Charcot's charismatic carriage and performances worked by raising expectations of improvement – the placebo effect.

Irrespective of the ultimate value of his work, Charcot inspired a generation of European intellectuals, a good example of which was the young philosophy teacher in Le Havre. Indeed, when a small ward was placed at Janet's disposal in Le Havre hospital, he called it – in jest, but nevertheless revealing a deep respect – Salle Saint-Charcot. Not only had Charcot's work suggested to the young Janet that hypnotism might be used to good effect in clinical settings, but it also suggested that 'physical' problems might have a psychological origin. Moreover, this psychological origin might take the form of 'ideas' that the patient was ordinarily unaware of. These fixed ideas – *idées fixes* – were perhaps lodged in some inaccessible part of the mind that could be exposed by hypnosis. This concept, one that Charcot never fully developed himself, made a deep impression on the young Janet. Indeed, its influence would affect the direction of his work for many years.

After the extraordinary (although ultimately unsatisfactory) case of Madame B, Janet was given permission by Dr Gibert to examine another young woman. Although her case would never lure the likes of Frederick Myers across the English Channel, it would inspire Janet to consider more carefully the clinical implications of his 'experiments'.

Lucie was only nineteen. She suffered from an anaesthetic (insensate) hand and hysterical paralysis of the arm. Moreover, she was tormented by fits of absolute terror and extreme anxiety. When asked to account for her behaviour she could give no explanation. Her perplexed and repeated response was merely, 'I don't know why'. In order to understand the origins of her fear, Janet employed a technique usually associated with the practice not of medicine but of spiritualism: automatic writing.

Janet had found that patients left holding a pen would begin writing automatically, particularly if their attention was focused elsewhere. He subsequently began to refine his technique. Janet enlisted the help of a colleague who conducted a conversation with Lucie. Meanwhile, Janet placed a pen in Lucie's anaesthetic hand and whispered questions to her from behind. Lucie seemed completely unaware of Janet's questions; however, without interrupting her primary conversation, she began to write. Although Lucie was ordinarily unable to tell her story, a part of her mind – dissociated from the rest – was able to write it.

When Lucie had been seven years old two men, concealed behind a curtain, had frightened her as a practical joke. When Lucie experienced her fits of terror, a part of her mind was reliving that same moment; however, the truly remarkable discovery, was that this 'part' of her mind had acquired a separate identity. Lucie was not only a nineteen-year-old woman, she was also Adrienne – a secondary personality fated to exist in the overwhelming and terror-filled world of a child. The early trauma had fragmented a single consciousness into two, one concealed within the other like a Russian doll. Janet hypnotised Lucie and, exploiting her state of heightened suggestibility, began to remove her symptoms. Adrienne disappeared. Janet observed: 'It is because the unconscious has its dream; it sees men behind the curtains, and puts the body in the attitude of terror.'

Although Janet's treatment method was not very different from his contemporaries (removing symptoms by suggestion in hypnotised subjects was not new), his investigative technique (automatic writing) was novel and inventive. Moreover, Janet was clearly already formulating a sophisticated theory. Traumatic memories could be replaying, over and over again, in the unconscious, literally paralysing the body with fear. He was definitely on to something.

Janet's next case provided him with a yet more powerful demonstration of the effects of unconscious thoughts and memories. A young woman, known as Marie, had been brought from the country to Le Havre hospital. At the age of only nineteen she was considered both insane and incurable. Marie suffered from convulsive 'hysterical' attacks with delirium, which subsequently lasted for days. After several months of observation it became clear that Marie's symptoms were associated with the onset of menstruation. She still suffered from psychiatric symptoms between menses; however, these were far less severe than the dramatic 'attacks'. Various parts of her body would go numb (for no reason) and she would experience episodes of 'terror'. In addition, Marie claimed to be completely blind in her left eye (although medical investigations suggested that her eye was in perfect condition).

Just before the onset of menstruation, Marie's character would suddenly change. She became gloomy, more violent, and exhibited nervous muscular spasms. However, the most curious symptom that developed was 'shivering'; as though the temperature had suddenly dropped. Twenty hours after her menstruation had begun, her menses would suddenly stop and a great 'tremor' would pass through her body. Marie would then experience a sharp pain, rising slowly from her abdomen to her throat. At this point, she would begin to convulse for a short period of time before experiencing a long and severe delirium.

While delirious, Marie would sometimes engage in the most disturbing behaviour. She would cry out, with an expression of terror on her face, and ceaselessly speak of blood and fire. Sometimes she would run, seeking desperately to escape imaginary flames. The hospital ward would be thrown into chaos as she mounted furniture or climbed on the stove. Marie's delirium and violent bodily contortions alternated with brief periods of rest for forty-eight hours. Then the 'attack' would finally end with a macabre finale, during which Marie would repeatedly vomit blood. With this extraordinary climax, the feverish episode of activity and delirium was brought to a close. After two days' rest, Marie would become quiet, and could remember nothing of her attack.

The doctors at Le Havre undertook traditional treatments. Marie was given medication and 'hydrotherapy' (a procedure involving immersion in water). Predictably, these were ineffective. After seven months of hospital residence (and seven dramatic attacks), the doctors were at a loss. Janet had been aware of Marie, but he was not eager to be involved. Any attempts at psychological intervention appeared to make Marie worse. For example, talking to her about her menstruation only made her delirium more extreme. Between attacks Janet had conducted a few hypnotic experiments on Marie, examining her anaesthesiae, but he remained a peripheral figure in her life.

After eight months, Janet happened to be undertaking one of his routine 'experiments' with Marie. She was 'between' menses, and therefore quite lucid. The young woman was very depressed, and she said to Janet, in despair, that she believed that she would never get better. She was doomed to a life in medical institutions. Perhaps Marie's sorry disclosure touched Janet. He asked her what it was that she experienced before getting sick. She replied: '… everything stops. I have a big shivering, and I don't know what happens next.' Janet asked her more questions, particularly about her menstruation; however, Marie was unable to give any clear answers. She became more and more vague and seemed to have forgotten almost everything that Janet thought might be relevant. Nevertheless, a human bond had been forged. Janet the philosophy teacher, Janet the experimental psychologist, was somehow more inclined to help this tormented young woman.

Because of his work with Lucie, Janet suspected that Marie's symptoms were caused by some inaccessible memory. A trauma perhaps, now lost below the threshold of consciousness. Janet hypnotised Marie and asked her questions. These were more or less the same questions he had asked while she was awake. This time, however, Marie answered. There was a story to be told. A story that she could only tell when asleep.

At the age of thirteen Marie had started to have her first period. She was filled with shame. Whether out of ignorance, or having been given misleading information, Marie was disgusted by what was happening to her. She had formed the belief that she might be able to stop the bleeding and her preferred method of doing this involved plunging herself into a bath of freezing water. Exactly twenty hours after the start of her first period Marie's plan succeeded. She was seized with a violent shivering and her menstruation stopped; however, shortly after, she became quite sick and for several days was in a delirious state. Eventually the delirium passed and Marie returned to normal health. She did not menstruate again for another five years. When she did, her shame and distress returned.

Janet summarised his findings in the following way:

Every month, the scene of the cold bath repeats itself, brings forth the stopping of the menstruation and a delirium which is, it is true, much more severe than previously, until a supplementary haemorrhage takes place through the stomach. But, in her normal state of consciousness, she knows nothing about all this, not even that the shivering is brought forth by the hallucination of cold. It is therefore probable that this scene takes place below consciousness, and from it the other disturbances erupt.

Like Lucie, Marie's severe psychological problems stemmed from a lost traumatic memory. Janet had the idea that he would rebuild her memory; plant in her mind, a 'false' memory. One that had no negative associations:

It was necessary to bring her back, through suggestion, to the age of thirteen, put her back in the initial circumstances of the delirium, convince her that the menstruation had lasted for three days and was not interrupted through any regrettable incident. Now, once this was done, the following menstruation came at the due point, and lasted for three days, without any pain, convulsion or delirium.

Janet had treated an incurable patient and he had done so using entirely psychological means. This was a remarkable breakthrough. Yet his understanding of Marie was not complete. He had not understood all of her symptoms. So he continued foraging in her unconscious mind for clues.

He discovered that Marie's episodes of terror were the repetition of an emotion first experienced when she was sixteen years old. Marie had witnessed the suicide of an old woman who had thrown herself down a flight of stairs. The 'blood' that she spoke of during her delirium was not menstrual blood but the old woman's blood. Janet treated Marie's terrors using his new technique. He hypnotised her and took her back to the time of the suicide. He told her that the old woman had merely stumbled; she did not intend to kill herself. Once the new memory was in place, the attacks of terror did not recur.

Janet then turned his attention to Marie's blindness. She was reluctant to discuss her impairment, and told him that she had been blind in her left eye from birth; however, Janet found that when he regressed Marie under hypnosis to the age of five she regained her ability to see. Clearly, something had happened when Marie was six years of age that had resulted in a form of 'hysterical' blindness.

The origins of Marie's blindness were eventually discovered in the form of what Janet described as a 'trifling incident'. Against her will, Marie had been forced to sleep with another child whose face was affected by impetigo on the left side. Impetigo is a bacterial skin infection which produces small fluid-filled blisters that eventually burst, leaving a brown crust. Some time later

Marie herself developed impetigo in exactly the same place. The impetigo was eventually treated; however, neither her parents nor her doctors noticed that she had developed a numbness (or deadening) affecting the left side of her face and that her left eye had lost the power of sight. Janet took his patient back through time again and began to rebuild her memories. Without drama, and in a rather matter-of-fact, almost inconsequential way, Janet recorded his therapeutic procedure and its successful consequence:

> … she caresses without fear the imaginary child. The sensitivity of the left
> eye reappears without difficulty, and when I wake her up, Marie sees clearly
> with the left eye.

Janet's treatment was complete. Where every other method known to medical science had failed, he had produced a miraculous transformation. This rather curt summary of his methods, is complemented by a near pathologically modest assessment of his own achievement:

> I do not attach to this cure more importance than it deserves, and do not
> know how long it will last, but I found this story interesting as showing the
> importance of fixed subconscious ideas and the role they play in certain
> physical illnesses, as well as in emotional illnesses.

The young philosophy teacher continued to publish his results in the *Revue Philosophique*. He investigated several more of Dr Gibert's patients and, as his work progressed, he become increasingly convinced that memories and ideas, outside of awareness, were the ultimate cause of mental illness. He coined his own term to describe this level of mind. He called it the *subconscious*; however, more importantly, he had demonstrated that hysterical symptoms could be cured, through the discovery and rebuilding of material in the subconscious. This process was much more sophisticated than what had gone before. Janet was not merely hypnotising patients and removing symptoms. Janet was uncovering and reshaping inaccessible memories.

Today, Janet's special term 'subconscious' is used in exactly the same way as the more commonly used 'unconscious'; however, Janet introduced the term subconscious for a very specific reason. He wished to make a clear distinction between the concept of the unconscious mind, as the source of mental illness, and the unconscious as described by philosophers – a more general concept still largely associated with romanticism. This was a useful distinction to make. Unfortunately, the term unconscious continued to be used indiscriminately by both doctors and philosophers.

Janet worked with patients in Le Havre from 1882 to 1888. Although he wrote several academic papers in which he recorded his experiments and findings, the ultimate destiny of much of this work was his thesis, on *psychological automatism*. The term psychological automatism was used to describe behaviour that occurs in the total or partial absence of awareness. At its most basic, psychological automatism could take the form of *absent-mindedness* – trivial actions executed largely while thinking about other things (for example, trying to put on a second pair of trousers). But psychological automatism could also take more complex forms. These forms of automatism were well documented among individuals suffering from a particular type of epileptic seizure. A patient might stop talking mid sentence and fail to respond to any verbal prompts, such as 'What's the matter?' The patient might then finish drinking a cup of coffee, rise, and walk to the door. At that point, he or she might recover consciousness and look around, somewhat confused, wondering what had just happened. In certain individuals these *absence automatisms* can be sustained for a considerable period of time. The patient's mind appears to be out of commission, but his or her body still manages to execute a limited repertoire of behaviours. Post-hypnotic suggestion and cases of multiple personality can be considered as even more complex examples of automatism. Both of these phenomena can be associated with the most sophisticated behaviours imaginable – for example holding a coherent and novel conversation – apparently without the consent (or participation) of the individual's conscious mind. In his thesis Janet catalogued numerous examples of psychological automatism and attempted to classify them.

Janet's most important work during his Le Havre years was, however, his exploration of hysteria. He had suggested that hysterical symptoms originated in parts of the mind that had been split off. These symptoms were, therefore, examples of psychological automatism. They were generated by parts of the mind that had become 'detached' and contained subconscious thoughts and memories. Janet believed that this 'splitting' was caused by the experience of trauma.

Janet did not include any of his experiments on telepathy in his thesis. A further notable omission was any reference to the possible therapeutic implications of his work. Janet was not, at that time, medically qualified, and he did not want to offend the sensitivities of the medical establishment. He was examined on 21 June 1889 at the Sorbonne. He passed without difficulty and was warmly congratulated for his achievement.

By 1889 Janet had completed the first major phase of his life's work. That August, he attended the International Congress for Experimental and Therapeutic Hypnotism in Paris. He was a committee member and well respected by his peers. Over 300 delegates attended. Many were distinguished. They included Frederick Myers, Cesare Lombroso (one of the first criminologists), and William James (the psychologist and brother of Henry). One member was yet to make his mark on the world. In fact, it would be another four years before he published his first important paper on the psychological treatment of mental illness. He was a young neurologist from Vienna called Sigmund Freud. The man history would canonise as the father of psychotherapy.

Janet began his medical studies in November 1889; however, his time was still not fully his own. He was still pledged to undertake teaching duties and for a year occupied a post at the Lycée Louis-le-Grand before moving to the Collège Rollin. Fortunately, he was regarded as a special case and he was able to spend a considerable amount of time on Charcot's wards at the Salpêtrière from 1890 onwards.

The Salpêtrière was, of course, a hospital. But to call it a hospital would fail to give a proper impression of its dimensions, for the Salpêtrière was not so much a large building designated for the care of the sick, as a city. Indeed, it was often described as a city within a city. The Salpêtrière housed thousands of patients (mostly women) in over forty buildings and covered over 125 acres of land.

Janet arrived at the Salpêtrière when Charcot's career was at its zenith. If the Salpêtrière was a kingdom in miniature, then there was no doubting the identity of the demiurge who presided over its shadowy and labyrinthine corridors. Charcot was known as 'The Napoleon of the neuroses', and his colleagues and acquaintances were called – by envious rivals – 'la charcoterie'. Janet had already been noticed and praised by Charcot, so when he arrived at the Salpêtrière his place among la charcoterie was virtually guaranteed. As soon as Janet commenced his medical studies he resumed his psychological investigations. At the Salpêtrière, he was allowed to assess and treat hysterical patients. Indeed, Charcot often referred his own patients. It is very likely that Charcot recognised that Janet was not really a student; at least, not in the conventional sense. He had gained a considerable amount of clinical experience while 'experimenting' at Le Havre and was, in his own way, an expert.

In his writings, Janet began to refine his treatment methods. He believed that treatment depended on a thorough understanding of psychological problems. Symptoms should be carefully analysed. Moreover, the clinician should attempt to reconstruct the progress and development of the illness. Symptoms concealed stories.

In order to facilitate the detection of such stories, Janet established a number of guidelines for good clinical practice. Today, these would be regarded as routine; however, for a practitioner in the 1880s they were extremely thorough and systematic. Janet suggested that the clinician should be alone with the patient at the time of examination. He believed that an exact record of everything said in therapeutic sessions should be recorded. Finally, and most importantly, the clinician should obtain an entire life history and consider the effects of any prior treatment. An adage of Janet's was that the clinician could never know enough about his patient. These recommendations show a strong inclination towards empiricism. Indeed, the collection of data, the formation of hypotheses – and experiment – were central to Janet's method.

One of the first patients that Janet saw at the Salpêtrière was Marcelle, a twenty-year-old woman whose problems had begun in early adolescence. Her principal symptom was paralysis, most notably, difficulty moving her legs; however, she also suffered from disturbances of memory and thought. Janet noticed that habitual movements were accomplished without too much difficulty, whereas movements that were preceded by voluntary decisions were extremely difficult. When, for example, Marcelle decided to get up and walk, her thoughts became disturbed by what she described as 'clouds': the entry into awareness of muddled ideas and even hallucinations.

Marcelle's memory problems were of an unusual nature. Her memory for events in her life up until the age of fifteen were good; from that age onwards her memories became intermittent or unclear. She could recall nothing of her life beyond the age of nineteen.

Janet began to piece together her personal history, which seemed to have considerable relevance to her symptoms. At the age of fourteen Marcelle had contracted typhoid. She became very ill and was unable to cope with life. Janet believed that at that point Marcelle had attempted to escape from her troubles by retreating into a world of daydreams and fantasy (a world, perhaps, of 'clouds'). This tendency may have been amplified by her voracious appetite for 'escapist' literature. The following year Marcelle's father – who had been paraplegic – died. The final telling event occurred when Marcelle was nineteen. An unsuccessful romantic involvement ended unhappily, and she became suicidal.

At first Janet tried to treat Marcelle by using the methods he had employed in Le Havre. He wanted to retrieve the ideas that Marcelle had formulated at various significant points in her life. To this end, he used hypnosis and automatic writing. Janet worked with the ideas and memories that became accessible; however, any improvement in Marcelle's condition was only temporary. As one symptom disappeared, another would take its place.

When Janet used hypnosis and automatic writing on Marcelle she experienced 'hysterical crises', becoming severely disturbed. Nevertheless, Janet made two important observations. Firstly, Marcelle appeared to feel better immediately after a hysterical crisis had been induced and, secondly, the more severe these crises became, the more they released ideas that were formed earlier in life.

Janet began to work with these ideas in reverse order, working back from the most recently formed ideas to the most remote in time. He wrote:

> By removing the superficial layer of the delusions, I favoured the appearance
> of old and tenacious fixed ideas which dwelt still at the bottom of her mind.

Once these older ideas were retrieved and modified, Marcelle's improvement was significant and lasting.

Janet's account suggests a number of interesting points. To treat a patient properly, it is necessary to retrieve and modify the most fundamental and relevant ideas and memories from the unconscious. Retrieval of unconscious material is best achieved if the patient is placed in a heightened emotional state. Moreover, the most important material in the mind is revealed, as it were, by excavation: the upper strata contain fewer riches than those laid down at an earlier time. Systematic improvement will be associated with the dissolution of pernicious ideas and memories of increasing priority. In the same case history, Janet wrote: 'in the human mind, nothing ever gets lost'.

Janet's reputation had begun to spread beyond the city within a city. The intellectual climate of the time favoured a general interest in the mind and its secrets. Inevitably, the literature of the day reflected this trend and began to explore psychological themes. Marcel Prevost, a famous contemporary writer, chose to make psychotherapy a principal feature of his 1893 novel *The Autumn of a Woman*. A central character in this work is Dr Daumier, a young neurologist practising at the Salpêtrière (who is also a gifted psychotherapist). The methods he employs to treat his patients borrow much from Janet. Moreover, his mannerisms and style of speech are believed to resemble Janet's. In all likelihood, Dr Daumier *is* Janet. Regardless of his retiring nature, the Salpêtrière constituted too small a bushel to conceal Janet's brilliance.

In the 1880s and 1890s becoming a doctor was a relatively undemanding business and, for Janet, required hardly any effort at all. On account of his prior academic training and experience, Janet was afforded many exemptions. Thus, he was able to sit his final examinations on 31 May 1893. He presented his MD thesis a few months later. Predictably, he graduated with honours. Moreover, Charcot (who was always interested in experimental psychology) created a laboratory at the Salpêtrière for the purpose of conducting psychological experiments. He gave it to Janet to run. In Charcot's eyes, Janet had clearly lived up to all of his early promise. A remarkable achievement, given that it was only eight years earlier that Janet's first major paper on Madame B was read at the Société de Psychologie Physiologique.

Janet's theory of hysteria was detailed in several academic journals and in his MD thesis (which was published in 1893). In summary, he suggested the following: Hysterical symptoms were due to the presence of thoughts and memories in the unconscious and have a traumatic origin. The traumatic experience is perceived as so overwhelming that it cannot be integrated into the psyche. Memories of the traumatic experience are subsequently 'split off' from awareness; sometimes to such a degree that months of a patient's life are consigned to oblivion.

This splitting of the mind is sometimes technically described as *dissociation*. However, these split-off or dissociated portions of mind can still exercise an influence on everyday existence; they 'return', to haunt the body, like ghosts, producing inexplicable impulses, anaesthesiae, and re-enactments. With respect to the latter, the individual is like an actor who is driven to perform a scene in the absence of props or other members of the cast. However, these re-enactments might not always be literal. They could, for example, be symbolic. For Janet, the so-called 'hysterical crisis' was almost invariably a re-enactment of some kind, albeit a disguised re-enactment. Keeping traumatic memories at bay requires mental energy. Janet believed that this produced what he described as 'a narrowing of consciousness'.

Janet used the term 'psychological analysis' to describe his methods. It is a term that has many contemporary variants, such as psychoanalysis or analytical psychology. However, these latter terms were employed to describe specific schools of psychotherapy associated with specific figureheads. Janet never attempted to form a school of psychotherapy. Moreover, he never used the term 'psychological analysis' with the intention of giving his approach a ready handle. He used the term as a simple and modest description of the process he used. That is to say, the methods he endorsed were geared towards *analysing* the *psychological* world of a given individual. A second stage, synthesis, would always be necessary if treatment was to be successful. Once the psychotherapist had established which parts of the mind had been 'split off' or dissociated, the task in hand would then be to reintegrate them.

Janet's work at Le Havre and the Salpêtrière was outstanding. He invented what we would now recognise as psychotherapy and linked his treatments to a specific model of psychopathology (i.e. a model of how symptoms develop).

Central to his model was what he called the subconscious. Without retrieving and modifying subconscious material, hysterical symptoms would persist.

Janet continued to investigate the mind and write about its mechanisms for the rest of his life. Just before he died he completed *The Great Synthesis* – a theoretical framework within which he believed all human behaviour could be understood. It consisted of over twenty volumes and dozens of scientific articles. In addition, he found the time to write some highly regarded books on philosophy. He also wrote on diverse subjects such as the evolution of moral and religious conduct, social behaviour, graphology, the paranormal, Francis Bacon, medical history, and criminology (to name but a few key areas). Janet was also something of a speculator. He believed, for example, that one day, men would invent the 'paleoscope' (or what H.G. Wells later called a time machine). He suggested that the world should learn a common language (an idea that eventually became esperanto), and predicted the use of powerful mind-altering drugs in clinical practice (what he then called 'narco-analysis' and what we would now call pharmacotherapy).

In most histories of psychology and medicine, Janet is usually mentioned in passing as a pupil of Charcot who studied hypnotism and experimented with it in the context of hysterical phenomena. His actual achievements – which were monumental – are almost entirely forgotten. This has prompted the Janet scholar Henri Ellenberger to write:

> ... Janet's work can be compared to a vast city buried beneath ashes, like
> Pompei. The fate of any buried city is uncertain. It may remain buried forever
> ... But it may also perhaps be unearthed some day and brought back to life.

To date, Janet's hidden city has yet to be properly excavated.

The year 1893 was an eventful one. Janet was awarded his MD thesis and Charcot, the Napoleon of the neuroses, the prince of science, died. He had been receiving hate mail for years predicting his imminent death from a heart attack. One of his colleagues – intimate enough to know of his angina pectoris – must have been very happy on 17 August. Janet had lost both a teacher and a powerful patron.

It was also in 1893 that an article titled 'On the psychical mechanisms of hysterical phenomena: preliminary communication' was published in Vienna. It is a paper of enormous historical importance in the annals of medicine and psychiatry. The 'Preliminary communication' is traditionally regarded as the first description of a psychological treatment of hysteria. It contains a passing reference to the work of 'the two Janets'. The authors were obviously familiar with the writings of not only Pierre but his less famous brother Jules; however, the general tenor of the piece is that the authors' only significant debt is to the Napoleon of the neuroses. In their concluding paragraph, their acknowledgement is unequivocal: ' ... we have taken a step forward along the path first traced so successfully by Charcot ...'

The 'Preliminary communication' was well received by many commentators all across Europe. Even Janet welcomed its publication; however, his cordiality was unmistakably tempered with a liberal helping of sang-froid. 'I am happy to see', wrote Janet, 'that the results of my already old findings have been recently confirmed by two German authors, Breuer and Freud.' Little did he realise that these two German authors were destined to take all the credit for discovering the role of the unconscious in mental illness. Moreover, one of them – Sigmund Freud – would become a figure of such enormous cultural significance that his overwhelming presence would eclipse Janet's life and work completely.

From the very beginning Freud was well aware of Janet's priority. But posterity was at stake. Subsequently, whenever Freud had the opportunity, he would describe misleading and distorted versions of Janet's ideas in his own works. He was tireless in his personal attacks on Janet, and even accused him of antisemitism.

In the April of 1937 Janet was passing through Vienna and requested a meeting with Freud. By that time, Freud was recognised the world over as the father of psychotherapy – the man who had unlocked the secrets of the mind. The man who had discovered the unconscious. They were both old men, and after a lifetime of enmity, much needed to be said.

Freud, the cultural icon, refused to see him.

4

The icon

On one of the hottest summer evenings of 1883, the young Sigmund Freud called on his friend and colleague Josef Breuer. In his dealings with Freud, the older man had come to assume the manner of a kindly uncle. Subsequently, when Breuer insisted that Freud should take a bath, his imposition was received with good humour. In fact, Freud later wrote to his fiancée, Martha Bernays, that Breuer had not only insisted that he take a bath, but Breuer had also chased him into the tub. It is clear from the tenor of Freud's letter that he did not consider Breuer's behaviour disrespectful, embarrassing, or improper. Clearly, as far as Freud was concerned, Breuer's avuncular conduct was perfectly acceptable.

When Freud finally emerged, having sloughed off the accumulated grime of the stifling day, he made his way upstairs to enjoy an evening meal with Breuer. Both men were at ease and sat in shirtsleeves, discussing medicine. In this atmosphere of relaxed bonhomie, the conversation drifted toward nervous problems and 'strange cases'.

Breuer had mentioned the name of Bertha Pappenheim before; however, on previous occasions, he had never divulged the details of her case in full. This was probably because Pappenheim was a friend of Martha Bernays and Breuer was no doubt concerned about breaching professional confidence; however, on this particular evening in 1883 Breuer decided to reveal all. But he was still uneasy, not confident that he was doing the right thing. He urged Freud not to discuss Pappenheim's history with his fiancée (or at least to refrain from doing so until they were married).

As night descended Breuer gave Freud an unexpurgated account of his involvement with Bertha Pappenheim. Hearing Breuer's story changed Freud's life and, as a consequence, the cultural climate of the twentieth century. As after-dinner conversations go, this was one that really mattered.

Bertha Pappenheim was an educated and creative young woman whom Breuer had accepted as a patient over two years earlier – in December 1880. Pappenheim's treatment had lasted for well over a year, so Breuer's memories of her (especially during the latter part of her treatment) were relatively fresh.

The first of Pappenheim's symptoms had appeared while she was nursing her father (who was suffering from an illness that eventually proved to be fatal). She was a devoted daughter and spent many testing hours by her father's sickbed. Indeed, the onset of her own illness had a negligible effect on her administrations. She persevered, caring for her father without complaint, but in due course the strain began to take its toll. Pappenheim's symptoms became more severe, and additional (rather florid) ones began to develop. Two months before her father's death she could no longer look after him. She had become too ill herself. Reluctantly, she relinquished her nursing duties.

When Breuer accepted Bertha Pappenheim for treatment she was already suffering from many symptoms that were assumed to have a hysterical (i.e. non-physiological) origin. These included headaches, loss of hearing, coughing, squinting, impaired vision, paralysis of several parts of her body, anaesthesiae, and general weakness. As her illness progressed, her symptoms became even more bizarre. She lost the ability to speak German and subsequently could only converse in English, French, or Italian. She suffered from extreme mood swings, exhibited a second (less respectful) personality, and stopped eating and drinking for extended periods. She even suffered from hallucinations. These tended to occur during the day and appeared to be linked by a supernatural theme. Thus, Pappenheim was frequently haunted by spectral black snakes, skulls, and skeletons – the macabre and terrifying trappings of an imaginary Gothic crypt.

In the evening, Pappenheim would enter an altered state of consciousness very similar to self-hypnosis, uttering the words 'tormenting, tormenting'. At this point, if a phrase she had spoken earlier in the day – while hallucinating – was repeated, she would begin to tell a story. Breuer visited Pappenheim every evening in order to listen to her stories (many of which reminded him of those written by Hans Christian Andersen). And every evening, when Pappenheim brought her story to a close, she would become calm, untroubled, and lucid.

Pappenheim was fully aware that this process was taking place, and subsequently called it 'the talking cure' (a term now used to describe all forms of psychotherapy). Somehow, allowing a deeper part of her mind to express itself was proving helpful. Cathartic, perhaps? But, unfortunately, the beneficial effects of the talking cure were short lived. Pappenheim's mental state reliably deteriorated over the next two or three days and she would once again find herself thrown into the same nightmare world of sinister hallucinations.

Breuer recognised that Pappenheim's self-induced trance state was of critical importance. It seemed to uncover a deeper stratum of Pappenheim's mind – the place where her therapeutic stories came from. Subsequently Breuer began to consider what other benefits might arise from probing this deeper level of mind. He decided to hypnotise Pappenheim, in order to exercise greater control over her entry into the trance state, and also to experiment further with the talking cure. Listening to Pappenheim's strange stories was all very well, but Breuer was now seized by the urge to be more proactive.

Once Breuer had hypnotised Pappenheim he asked her many questions about the nature of her symptoms. For example, he would take a symptom such as her squint or macropsia (seeing things as bigger than they actually are), and try to work out when it first appeared. This line of questioning led to an interesting discovery. Each of Pappenheim's symptoms could be traced back to a forgotten event that had occurred while she was nursing her father.

> For instance, on one occasion, when she was sitting by her father's bedside
> with tears in her eyes, he suddenly asked her what time it was. She could not
> see clearly; she made a great effort, and brought her watch near to her eyes.
> The face of the watch now seemed very big – thus accounting for her
> macropsia and convergent squint.

As Breuer recovered each of these memories, the symptoms that they were associated with disappeared. Pain and sadness, trapped in some inaccessible region of the mind, had been released. It appeared that Pappenheim's stories had been related to the operation of some spontaneous cathartic mechanism in the mind. The unhappy emotion attached to inaccessible, traumatic memories was being channelled into her melancholy narratives.

By using hypnosis Breuer had managed this process more effectively. He had devised a method of syphoning off unhappy emotions from their origin – unconscious memories.

The story of Breuer's treatment of Bertha Pappenheim made a deep impression on Freud. And even though the next three years proved to be very eventful, Freud never forgot Breuer's after-dinner disclosures.

In 1885 Freud went to Paris to study under Charcot at the Salpêtrière. There, seated among the doctors and dignitaries who attended Charcot's demonstrations, Freud watched as the prince of science created and removed hysterical symptoms using hypnosis and the power of suggestion. And as he watched, memories of Breuer's remarkable case stirred. Eager to elicit an opinion from the high priest of neurology, Freud approached Charcot and told him about Breuer's work. But Charcot was singularly unimpressed. He responded with indifference.

Yet this did nothing to dampen Freud's conviction that Breuer had stumbled across something interesting. Even Charcot's indifference failed to smother Freud's nascent plan to involve Breuer in some collaborative research.

In 1886 Freud returned to Vienna, married Martha Bernays, and went into private practice as a consultant in nervous diseases. He contacted Breuer and the two men began working together – discussing their patients and building on the slight theoretical foundation established by Breuer while working with Pappenheim. Patients with hysteria were hypnotised, traumatic memories were recovered, and improvements followed.

In 1893 Breuer and Freud published their short paper 'On the psychical mechanism of hysterical phenomena: preliminary communication', which eventually became the opening chapter of *Studies on Hysteria* (1895), a comprehensive account of their pioneering treatment methods. The first and most important case cited in this historically significant work, was the case of Bertha Pappenheim – renamed Fräulein Anna O in order to protect her identity.

In the 'Preliminary communication' Breuer and Freud did not use the term 'unconscious', but its existence is definitely implied. They suggested that hidden memories of certain experiences were the root cause of hysterical symptoms; however, they somehow managed to get by without using the term 'unconscious memories'.

> … these memories, unlike other memories of their past lives, are not at the patients' disposal. On the contrary, these experiences are completely absent when they are in a normal psychical state, or are only present in a highly summary form.

By the time Breuer and Freud came to write *Studies on Hysteria* the unconscious had become an essential concept in their theoretical framework. *Studies on Hysteria* contains an important chapter titled 'Unconscious ideas and ideas inadmissible to consciousness – splitting of the mind'. However, it was written entirely by Breuer, suggesting that, of the two authors, Breuer was considered the greater authority on the unconscious. Indeed, Freud's very first reference to the unconscious is rather inconsequential. It appears buried in a footnote attached to one of the case studies, that of Frau Emmy von N, a name rarely given much significance in the annals of psychiatry.

Studies on Hysteria merits more than a superficial reading. It tries to keep a secret – and fails. It tries to fudge an uncomfortable truth, but in doing so attracts further attention to a glaring inconsistency.

In his account of the case of Anna O, Breuer admits to 'having suppressed a large number of quite interesting details'. This statement appears to be a casual remark and Breuer immediately goes on to imply that some ruthless editing was necessary in order to contain an already much extended case study. His admission that some interesting details were omitted, however, follows a curiously ambiguous passage in which Breuer summarises the very final stages of Pappenheim's treatment.

> She was moreover free from the innumerable disturbances which she had previously exhibited. After this she left Vienna and travelled for a while; but it was a considerable time before she regained her mental balance entirely.

What does this mean? On the one hand, Pappenheim's treatment was successful and she was free of disturbances. But on the other hand, she was still mentally unbalanced.

The fact of the matter was that Breuer never really finished treating Pappenheim. In the summer of 1882 Breuer referred her on to a colleague, Dr Robert Binswanger, who cared for her at a Swiss sanatorium. The talking cure was abruptly terminated, and although Pappenheim's health had improved, she was still complaining of several symptoms; for example, odd absences of awareness, language problems, and psychogenic pain. Breuer had simply glossed over the ending of what in every other respect was a meticulously written case study.

Why? What went wrong?

Breuer and Pappenheim had developed a very close relationship. For over a year Breuer had been visiting her. He had sat by her side, evening after evening, listening to her poetic flights of imagination. She must have seemed like a nineteenth-century Scheherazade – captivating Breuer with her sad, whimsical stories. She fascinated him. Moreover, Breuer came to respect and admire her. When he described her he wrote of a woman who was imaginative, gifted, intelligent, 'sharp', and kind. He genuinely liked her.

In such circumstances Breuer was understandably shaken when, close to completing therapy, he discovered her one evening in a confused state, writhing around and clutching her stomach. When he asked her what was wrong she told him she was pregnant with his child.

This wasn't what he wanted to hear. Breuer was a respectable middle-aged doctor. He had spent a lot of time alone with a woman half his age, and this sort of allegation was most unwelcome. And, of course, there was also the matter of Frau Breuer to consider. What would she have to say?

Even though Pappenheim's symptoms were just about under control,

Breuer decided he couldn't carry on. It was all too risky. He packed Pappenheim off to Switzerland and tried to forget about the whole affair. That is, until a stifling evening in 1883 when he decided to disclose the whole story to his young colleague, Sigmund Freud.

Clearly, Breuer was still embarrassed by Pappenheim's phantom pregnancy when he wrote up the case of Anna O some twelve years after her 'treatment', hence his guilty admission of having suppressed certain details. However, some commentators – fancifully, perhaps – have suggested that even though his narrative was guarded, Breuer was unable to conceal his sexual desire for Pappenheim. By choosing the name Anna O to disguise Pappenheim's real identity he gave himself away. Feminist writers have pointed out that the circle, or letter 'O', is an ancient symbol representing the female genitalia.

If Breuer had completed Pappenheim's treatment in 1882 and then published her case study, he would have preceded Janet by several years. Although Breuer underscored some differences between his own and Janet's understanding of hysteria, such differences are relatively minor compared with their commonalties. Both men agreed that the symptoms of hysteria were caused by unconscious traumatic memories, and that the recovery of these memories was important if treatment were to be successful. The two men also share the dubious distinction of having been largely forgotten because of Freud's celebrity.

Breuer was a real pioneer. He was one of the first recognisably modern practitioners (as opposed to the early hypnotists) to devise a psychological treatment based on the recovery of unconscious memories. He also wrote the first extended essay on the unconscious in the psychoanalytic canon. Yet, after the publication of *Studies on Hysteria*, he played no significant role in the development of what became psychoanalysis and did nothing more to further contemporary understanding of the unconscious. He simply let Freud carry on without him.

The reason for Breuer's retreat from the avant-garde of nineteenth-century psychology was remarkably straightforward. It was sex – or at least his junior partner's growing obsession with it.

Freud became increasingly convinced that sexual feelings were very significant in the development of mental illness. Thus, he was gradually edging towards a sexual theory of hysteria. This was by no means an original idea. Freud was familiar with the work of another Viennese physician called Moritz Benedickt, who from 1864 had been suggesting that hysteria might be caused by a closely kept sexual secret, and that the symptoms of hysteria would disappear if such a secret were revealed. Therefore, by the time Freud was advocating his own sexual theory of hysteria this kind of thinking had already achieved some currency among broad-minded Viennese doctors. Unfortunately, Breuer wasn't one of them. Very swiftly an irreconcilable rift opened between the two erstwhile collaborators.

In the end, the case of Anna O probably divided Freud and Breuer more than it ever brought them together. Whereas Freud was extremely interested in the erotic implications of the case, Breuer found them awkward and embarrassing. It seems likely that whenever Freud insisted on the importance of sexual impulses in the development of hysteria, Breuer was rudely reminded of an intelligent and attractive young woman, clutching her abdomen, claiming to be in labour with his child.

Breuer had lifted the lid on the unconscious and forbidden fruit had tumbled out. But while Breuer was struggling to shut it all up again, Freud was already trying to step inside.

In 1897 Freud decided to see a psychoanalyst; however, given that he was the only one practising in Vienna at the time, he was more or less forced to consult himself. In a bizarre manoeuvre that still seems somewhat surreal, he put himself on the couch and attempted to storm his own unconscious.

This bold gesture is often portrayed by psychoanalytic hagiographers as an epic journey. Freud, like a Homeric hero, sets out on a perilous voyage of discovery. And when he returns, he does so as a changed man – altered by his experiences and in possession of arcane knowledge.

Of course, Freud was not the first to emulate the mythic heroes of Greece and Rome by descending into the underworld of his own unconscious. As we have already discovered, writers such as Coleridge and De Quincey had undertaken similar voyages, returning to document their visions and strange encounters in works of poetry and prose. Coleridge and De Quincey had experienced the unconscious directly in their dreams, which were made more colourful (and nightmarish) through the action of opium on the brain. The relationship between dreams and the unconscious did not escape Freud (who also shared with Coleridge and De Quincey a weakness for drugs in the form of a cocaine addiction). Freud recognised that by examining the contents of his own dreams he too might learn more about the workings of the unconscious mind. After analysing his dreams (and those of his patients), he was able to write a book that has since become a cultural landmark, *The Interpretation of Dreams*. Later, reflecting on his achievement, he wrote:

> The interpretation of dreams is in fact the royal road to a knowledge of the unconscious; it is the securest foundation of psychoanalysis.

The Interpretation of Dreams was published in 1899, but dated 1900. Only

600 copies were printed and it took some eight years to sell them all. Since then it has been a permanent presence in book shops all over the world and continues to sell in respectable numbers. The success of *The Interpretation of Dreams* can be ascribed to both its literary and scientific appeal. Indeed, some have argued that it represents an entirely new form of writing – a complex amalgam of scientific history, clinical case study, philosophical investigation, and candid autobiography. Although Freud continued to write about psychoanalysis for the rest of his life, he always insisted that *The Interpretation of Dreams* contained his most important work. In it, Freud established the unconscious as a central concept in his theoretical framework, explained the mechanisms that shape dreams, and justified his break with Breuer by 'proving' that repressed sexual urges (including those dating back to childhood) were causally related to the symptoms of mental illness.

The most striking feature of *The Interpretation of Dreams*, however, is the great significance that Freud gives to unconscious processes, not merely in relation to psychopathology and dream interpretation, but in the broader context of mental life. Already Freud was suggesting that the importance of consciousness was routinely overestimated, and that the really significant features of the human mind operated below the awareness threshold.

> The unconscious is the larger sphere, which includes within it the smaller sphere of the conscious. Everything conscious has an unconscious preliminary stage; whereas what is unconscious may remain at that stage and nevertheless claim to be regarded as having the full value of a psychical process. The unconscious is the true psychical reality; in its innermost nature it is as much unknown to us as the reality of the external world, and it is as incompletely presented by the data of consciousness as is the external world by the communication of our sense organs.

For Freud, then, the unconscious was 'the true psychical reality'. The great 'unknown'. One way or another, he spent the rest of his life either thinking or writing about it.

Freud described psychoanalysis in many ways; as a 'medical treatment', a specialist science, a branch of psychology, a depth-psychology, and a 'psychology of the unconscious' (to cite just a few examples). He wrote approximately three and a half million words on psychoanalysis and revised, modified, and developed psychoanalytic ideas throughout his life. Nevertheless, a number of key concepts survived these revisions, and they represent the foundations of psychoanalysis.

Freud's understanding of the fundamental forces operating in the mind was influenced considerably by nineteenth-century science; in particular, by the branch of physics known as dynamics. Observations had shown that forces can combine or work in opposition; energy can be generated, stored, and eventually discharged. Freud believed that the human mind also contained energies, and that these behaved in much the same way. For Freud, the most important energy in the mind (and body) was sexual. He called this *libido*. According to Freud, the build-up of sexual energy in the body is associated with the intrusion of sexual thoughts and images in awareness. The ensuing state of tension is ordinarily reduced by sex itself.

According to Freud, mental life takes place on three different levels. He described them as *conscious*, *preconscious*, and *unconscious*. The standard method of explaining their relationship is the searchlight analogy. If the beam of a searchlight represents awareness, then objects illuminated by that beam reflect the contents of the conscious mind. Some objects will be just outside the beam but within its range. Those can be termed preconscious; however, some objects are simply too far away to be illuminated, and these correspond with the contents of the unconscious.

Much later, Freud also divided the mind into three 'agencies'. The *id*, *ego*, and *superego*. They are not simply different parts of the mind, but powers – each with a specific function.

The id operates unconsciously, and its principal function is to bias mental life in such a way as to engineer the discharge of libido. The ego has conscious, preconscious, and unconscious regions, and it has a largely managerial function – initiating and monitoring thoughts and movements. The superego too has conscious, preconscious, and unconscious regions, but its function is to act as moral judge. In everyday language, the id can be thought of as the source of passion, the ego can be thought of as 'the self', and the superego as corresponding with the idea of conscience.

When babies are born, their behaviour is entirely dictated by the id. Life is nothing but striving to satisfy basic needs and drives that are shaped, at least initially, by unconscious processes. Among these drives is a primitive form of the sexual instinct, motivating the baby to seek the 'sensual' (or at least physical) pleasure of maternal contact. With experience, a section of the id separates, and becomes organised around an identity – the ego. Later in infancy, the absorbed values and attitudes of family and society achieve a degree of independence, and the superego is formed; however, it should not be forgotten that parts of both the ego and the superego remain unconscious. Although the human mind develops self-awareness, no division of the mind is completely transparent. Thus, in Freud's scheme, the unconscious is relevant to every level and division of the mind.

In *Phaedrus*, Plato suggests that the soul, or mind, is like a chariot drawn by two very different horses. One is noble and refined while the other is base and impulsive. To steer the chariot, the charioteer must take into account the temperament of both horses. This is exactly the position of the ego in Freud's three-way division of mind. The id is constantly seeking to satisfy libidinal urges while the superego is constantly holding back. To steer a steady course, the ego must evaluate each situation in turn, and balance the id's instinctual needs against the superego's capacity for disapproval.

Because the id and superego make conflicting demands, a compromise must be reached. This compromise is accomplished by the ego, when defence mechanisms are activated. The most common defence mechanism is *repression*. If an unacceptable thought or wish enters awareness, this will produce discomfort in the form of anxiety. The offending thought or wish is then automatically forced below the awareness threshold, and anxiety subsequently subsides. An individual does not make a conscious decision to repress material. The mechanism functions unconsciously.

Unwanted thoughts and wishes may not always cause discomfort, and the psychic ban may be lifted in due course. Thus, defence mechanisms work like a psychological thermostat, maintaining a state of mental equilibrium.

Unfortunately, there is a cost attached to the deployment of defences. Energy is required to maintain a defence, leaving less energy available for other uses such as general day-to-day functioning. The situation is analogous to an advancing army establishing several defensive garrisons but at the same time reducing its overall power. An ego that is protected by too many defences may be too weak to function. When this happens, the affected individual is likely to complain of symptoms associated with mental illness.

Freud specified two principal situations necessitating the deployment of costly defences. First, when an individual has had a traumatic life experience (particularly an early one) and second, when an individual has failed to mature properly, and is subsequently revisited by the inappropriate wishes and desires of infancy. The most celebrated example of the latter is the Oedipus complex – in which the strong (and often sensual) attachment to mother fails to resolve itself in adolescence and survives into adulthood. The Oedipus complex is also associated with hostile feelings directed at the father (on account of his privileged access to the mother). In both cases, either traumatic memories or unacceptable Oedipal wishes must be kept out of awareness to reduce discomfort. Keeping such material in the unconscious consumes a considerable amount of psychic energy.

Although repression is the most common defence, Freud discussed many others, though he did not attempt to describe them in great detail. This task of naming, cataloguing, and defining defences was undertaken mainly by his daughter Anna. Examples include *reaction formation* (in which real feelings are masked by an intensification of their opposite) and *projection* (in which unacceptable thoughts or feelings are attributed to others). Like repression, defence mechanisms such as reaction formation and projection are deployed unconsciously. Thus, the individual does not realise the extent to which reality is being distorted. Freudian psychotherapy is largely concerned with the achievement of insight and the correction of distorted perceptions.

One of Freud's most enduring analogies is his comparison of therapy to the Dutch land-reclamation project. He suggested that the ego was like a land mass threatened by the rising waters of the id. Familiar landmarks slowly disappear as thoughts, feelings, and behaviour are increasingly determined by unconscious processes. In the same way that the Dutch reclaimed the Zuider Zee, so must the psychoanalyst win back parts of the mind that have succumbed to the unconscious. When psychoanalytic treatment is successful, the individual can abandon redundant defences, thus liberating sufficient energy to strengthen the ego. The individual can live a full life, in touch with the present moment and free from distortions.

To achieve therapeutic results, the patient must gain insight into his or her symptoms; however, such insight is difficult to accomplish – the major determinants of symptoms (such as repressed memories) and the operation of defence mechanisms being unconscious. Even so, although the unconscious guards its secrets closely, it is not beyond betraying itself. It sometimes makes mistakes or 'leaks' information. Subsequently the observant analyst is able to recognise significant signs and interpret them for the patient. Like Breuer, Freud's first attempts at revealing the content of the unconscious involved hypnosis. But he found the technique unsatisfactory. For example, one of his early patients – Fräulein Elisabeth von R – responded to his mesmeric efforts by proclaiming triumphantly: 'I'm not asleep you know; I can't be hypnotised'. Subsequently Freud soon developed his own methods, and by rejecting hypnosis managed simultaneously to distance himself from a shamanistic tradition and enhance his own scientific credibility.

Freud discovered that if patients were encouraged to talk freely, without constraint, they tended to touch upon subjects and experiences highly relevant to their symptoms. He called the technique *free association*. Freud also noticed that the unconscious gave itself away through verbal errors – the so called 'Freudian slip'. Careful consideration of such slips revealed the patient's true (and often inappropriate) intentions and wishes. Freud also suggested that the relationship between therapist and patient could reveal important information. Often, patients re-enacted elements of previous meaningful relationships, unconsciously projecting the characteristics of parents or significant others on to the therapist. He called this phenomenon *transference*; however, the most important method of probing the unconscious was the 'royal road' of dream interpretation.

Like many commentators before him, Freud believed that, in sleep, the barrier between the conscious mind and unconscious mind was weakened. As a result, urges and desires usually confined in the unconscious could escape and ascend into awareness – posing a threat to sleep (as the sleeper could easily be shocked into a waking state by unacceptable images). Because sleep has a restorative function, repeated awakening might eventually injure health. Consequently a censor operates in the mind, transforming unacceptable material into acceptable material that will not disturb the sleeper.

Freud distinguished between the *manifest content* and *latent content* of dreams. The manifest content is the edited or transformed dream, the latent content is the unedited or untransformed dream. The reason why dreams often don't make very much sense, is that the original narrative has been subject to so many transformations. The purpose of psychoanalytic dream interpretation is to decode the manifest content to reveal the underlying latent content. This is achieved by taking each element of the dream – be it image, thought, or feeling – and using it to initiate a chain of free associations. Predictably, the patient's associations touch on clinically relevant material and the psychoanalyst can infer the latent meaning of the dream when the process is complete.

Freud called the process by which unacceptable material is disguised *dream work*. Mechanisms operate in the sleeping mind that perform the same anxiety-reducing function as defences when the mind is awake. Freud described several of these mechanisms, but the most well known (and easiest to grasp) is *symbolisation*. When symbolisation occurs, an unacceptable object is transformed into an acceptable alternative. Like von Schubert some forty years earlier, Freud saw phallic significance in knives and clarinets when they appeared in the context of a dream.

Although Freud never insisted that all dreams are ultimately sexual in nature, his clinical work suggested that the unacceptable or latent content of dreams was frequently associated with a forbidden erotic wish. Indeed, he believed that some kind of *wish fulfilment* was the most unvarying feature of all dreams.

Clearly, Freud's theoretical framework and treatment methods place enormous emphasis on the role of unconscious memories, processes, and motivation. In 1915 Freud started to write a series of papers on metapsychology. He subsequently produced a collection of complex, densely written articles,

containing some of his most significant theoretical formulations. It is generally agreed that the most important of these is a seven-part essay on 'The
Unconscious'.

Freud was always at his most combative when defending the concept of
the unconscious – an activity that he repeatedly returned to throughout his
career. Remarkably, he was still justifying the concept when he died. An unfinished scrap of theoretical writing was discovered after his death which contained yet another defence of the unconscious.

Why was Freud so indefatigable in this respect?

Obviously, Freud was keen to strengthen the theoretical foundations of
psychoanalysis. The entire edifice was underpinned by the unconscious, so it
was essential that these subterranean struts and supports were well maintained, but Freud was also simply responding to criticism. Although the
unconscious had been mooted since the time of Leibniz, it was still regarded
as a controversial concept. Many philosophers, scientists, and doctors continued to approach the unconscious with scepticism and a significant body of
academics rejected it altogether. There were even dissenting voices within
the psychoanalytic movement. Several heretical disciples believed that Freud
had exaggerated the role of the unconscious and that more consideration
should be given to the accessible contents of the ego.

Resistance (and particularly academic resistance) to the unconscious is –
on the face of it – quite understandable. The unconscious is a concept that
balances precariously on the cusp dividing science from the paranormal. It is
invisible and intangible, and like most things that cannot be directly observed
or touched elicits mistrust. Human beings have a strong sense of self, possession of a unique identity that transacts with the world by making rational
choices. Although Freud's assertion that individuals suffering from mental
illness might show compromised insight and deficiencies of judgement was
plausible, his idea that all of humanity (to a greater or lesser extent) suffered
from exactly the same problem was construed by many as an offensive over-
generalisation. Most people don't feel out of control, or driven by dark, primitive forces. How could it be that almost all of human behaviour was
determined by unconscious processes? Freud's suggestion sounded preposterous.

Given Freud's tireless defence of the unconscious, it is surprising that initially he was rather cautious about the idea. The 'Preliminary communication'
does not refer to unconscious memories, but this is not because of ignorance
or the failure of the medical lexicon to supply an appropriate term. Freud was
already very familiar with the concept of the unconscious – and had been for
most of his life. In Freud's school were textbooks which described the work

of Herbart, and when Freud was studying medicine at Vienna seniors such as the physiologist Ewald Herring recognised 'unconscious material nerve-processes' and 'unconscious trains of ideas'. Freud would have encountered the unconscious in the context of German romantic philosophy and he was, of course, well acquainted with the work of Charcot and Janet.

Freud was a neurologist. Subsequently, when he began to develop an interest in psychological problems, he was still thinking like a neurologist. At the time of his early publications with Breuer his ultimate ambition was to give an account of psychological phenomena by describing the activity of brain cells and the flow of chemical energies. He was trying to do this by using physical rather than psychical explanatory concepts. This proved to be extremely difficult and by 1899, with the publication of *The Interpretation of Dreams*, he had completely given up trying. Describing mental life in terms of neuronal systems had proved inadequate. The subtleties of mental life could not be captured using a reductionist, physiological theory, and Freud was forced to adopt a psychological vocabulary – making use of the term 'unconscious' inescapable.

Although Freud cannot be credited with discovering the unconscious, he can certainly be credited with investing the greatest amount of energy into trying to understand it. More than any of his forerunners and contemporaries, he tried to show what the unconscious was like: how it worked and how it was different from the conscious mind.

Freud opens his 1915 essay on the unconscious with guns blazing – ready to address the critics:

> Our right to assume the existence of something mental that is unconscious
> and to employ that assumption for the purposes of scientific work is disputed
> in many quarters. To this we can reply that our assumption of the
> unconscious is necessary and legitimate, and that we possess numerous
> proofs of its existence.

Freud begins by addressing some very basic objections to the idea of the unconscious. The first of these is logical absurdity. How is it possible to arrive at knowledge of the unconscious? If material in the mind is really unconscious, then it is unknowable. Moreover, if unconscious material does enter awareness – becoming known – then it cannot be described as unconscious any more. To get around this problem Freud invokes the notions of transformation. The contents of the unconscious can be inferred by examining phenomena such as slips of the tongue, dreams, and symptoms. Unconscious material remains unconscious, but enters awareness in a modified form or distorts the contents of awareness from below the awareness threshold. In this way Freud neatly disposes of the apparent contradiction of knowing the unknowable.

Freud then appeals to the reader's everyday experience of mental life, which seems to depend on an unconscious substratum of activity. Freud notes that ideas (and sometimes solutions to problems) seem to just pop into awareness apropos of nothing. Where do they come from? Where are such problems worked out? Simple self-observation seems to suggest that many ideas are the result of prior stages of preconscious analysis or contemplation. Even ordinary thoughts and images just arrive in the stream of consciousness. We are not aware of their assembly, only their delivery.

In his unremitting defence of the unconscious, Freud then revisits a point originally raised by Augustine; namely, that an individual's totality must necessarily exceed the contents of his or her own awareness:

> We can go further and argue, in support of there being an unconscious psychical state, that at any given moment consciousness includes only a small content, so that the greater part of what we call conscious knowledge must in any case be for very considerable periods of time in a state of latency, that is to say, of being psychically unconscious. When all our latent memories are taken into consideration it becomes totally incomprehensible how the existence of the unconscious can be denied.

This is an important and very persuasive argument. Nevertheless, it is an argument that was considered by some to have a fundamental weakness: What if latent memories are not psychological but physical in nature? As Freud explains, latent memories might 'correspond to residues of somatic processes from which what is psychical can once more arise'. As such, if one adopts this position, there is no need to posit a psychology of the unconscious – a living, breathing world in which dynamic energies animate the dead record of memory. There is merely a neurochemical storage facility. A biological lending library, the volumes of which only become psychologically real when issued to consciousness.

Freud suggests that this approach offers few prospects for advancement. Indeed, it is something of an intellectual cul-de-sac. He argues that the physical characteristics of memory are totally inaccessible; for example, you can't see the memory of a sunny day or moonlit night by examining a slice of brain tissue under a microscope. Therefore, this form of criticism doesn't lead anywhere. It presents an alternative to a psychological unconscious, but then immediately runs up against the limits of scientific enquiry. If, on the other hand, one accepts the existence of a psychological unconscious, research into the mind and its working can continue in a meaningful way. In essence, psychoanalysis is a more practicable theory. Freud was not against biological accounts of memory. It was simply that thinking about latent memories as existing in a psychological unconscious (rather than as physical 'residues') offered richer theoretical possibilities.

Finally, Freud concludes his defence of the unconscious by reminding the reader that, prior to the advent of psychoanalysis, evidence for the unconscious had already been gathered by studies using post-hypnotic suggestion.

The subsequent sections of 'The unconscious' are highly technical. One senses that Freud is really trying to come to terms with the machinery of the unconscious. He is trying to dismantle it – metaphorically plunging his hands into the darkness and loosening nuts and bolts; feeling for parts; probing the gears; warming his palms on the dynamo.

Freud reiterates his early tripartite topography of mind – consisting of conscious, unconscious, and preconscious domains – and then seeks to define the special characteristics of the unconscious.

Freud states that the unconscious is exempt from mutual contradiction. It is a place where love and hate can comfortably exist, side by side. Like a machine designed by Escher, its impossible gears are not thwarted by logical inconsistencies. They smoothly work around stark juxtapositions and polarities.

The unconscious is timeless. Events are not ordered chronologically in the unconscious, nor are they altered by the passage of time. The recollections of early childhood are as potent as the recollections of the previous day.

Finally, the unconscious is a 'place' – a location where external reality has been replaced by what Freud called 'psychical reality'. It is a kind of psychoanalytic cyberspace. An inner landscape where virtually anything can happen. It has its own enigmatic truths. The psychological truths of fantasy and the dream world.

The end of the nineteenth century, and the beginning of the twentieth century, were important times in the world of archaeology. The German millionaire Heinrich Schliemann, guided by a careful study of Homer, had excavated Troy and discovered King Priam's Palace. He had excavated Golden Mycenae and found the tombs of Agamemnon, Cassandra, and Eurymedon. In 1900 Sir Arthur Evans excavated Knossos – site of the fabled labyrinth wherein Theseus slew the Minotaur. And, shortly after, the Golden Age of Egyptian archaeology began, culminating with Howard Carter's discovery of the tomb of King Tutankhamen in 1922.

Freud had a passion for archaeology. Indeed, he claimed to have read more archaeology than psychology and followed the fortunes of contemporary excavations with great interest. Moreover, his consulting room was crowded with ancient artefacts. His patients commented that the ambience he created had more in common with a museum or an archaeologist's study than a doctor's consulting room. Everywhere were ranks of statuettes and figures, fragments of eroded stone, glass cabinets displaying antiquarian treasures, and on his walls hung images of temples, sphinxes and pyramids; a young woman, stepping silently through the streets of doomed Pompeii. And among it all the famous couch, covered in pillows and an ornate Persian rug.

The analogy that can be drawn between psychoanalysis and excavation is obvious, and it is indeed interesting how the major discoveries of one discipline coincided with the other. Although the analogy between psychoanalysis and archaeology is a crude one, it was an analogy that Freud enjoyed. He believed it to be particularly apposite. While he was treating a famous case known as the 'rat man' (published in 1909 as 'Notes upon a case of obsessional neurosis'), Freud used the relics in his consulting room to make a point concerning the preservation of unconscious memories:

> I then made some short observations upon the psychological differences
> between the conscious and the unconscious, and upon the fact that
> everything conscious was subject to a process of wearing-away, while what
> was unconscious was relatively unchangeable; and I illustrated my remarks by
> pointing to the antiques standing about in my room. They were, in fact, I said,
> only objects found in a tomb, and their burial had been their preservation:
> the destruction of Pompeii was only beginning now that it had been dug up.

Freud saw himself as the mind's archaeologist; he had taken a spade and plunged it into the topsoil of the limen. Like Schliemann, Evans, or Carter, he was in the business of opening secret chambers, recovering relics, and translating hieroglyphs (although in his case these corresponded with the unconscious, repressed memories, and dream interpretation). It was inevitable, therefore, that Freud would eventually speculate on the existence of genuine ancient treasure, buried deep in the unconscious. Unable to participate in a 'real' archaeological dig, he began to contemplate the possibility of undertaking the nearest psychological equivalent.

Intimations of Freud's intention can be found in early letters to his friend and colleague Wilhelm Fliess. In 1897 Freud was using terms such as 'endopsychic myth' and 'psychomythology'. And in a letter dated 4 July 1901 he wrote:

Have you read that the English have excavated an old palace in Crete (Knossos) which they declare is the authentic labyrinth of Minos? Zeus seems originally to have been a Bull. It seems, too, that our own old God, before the sublimation instigated by the Persians took place, was also worshipped as a bull. That provides food for all sorts of thoughts which it is not yet time to set down on paper.

In this passage, one has a strong sense of Freud teetering on the edge of insight, daring to entertain 'all sorts of thoughts' which he was reluctant to share. But what were these thoughts? One can only speculate, but given the eventual direction of Freud's writing and his reference to Knossos and the labyrinth, it is relatively easy to hazard a guess. If the labyrinth was real, then perhaps myths were vestigial memories. In the same way that personal memories are preserved in the unconscious, might not cultural memories also be preserved in the unconscious – but at a much lower level?

These considerations eventually resulted in the publication of *Totem and Taboo* (1912–13), which is very probably Freud's most fanciful work. In it, he suggested that correspondences might exist between the Oedipus complex, the emergence of Oedipal themes in art and drama, and an actual prehistoric event in which the dominant male (or father) was murdered by the younger male members of a primal horde (or tribe) in order to gain access to females who had hitherto been jealously guarded. This was an extraordinary suggestion. Moreover, it carried with it the strong implication that ancient memories (and their associated emotions) could be passed down through subsequent generations. In theory, the memories buried at the deepest levels of the unconscious were far, far more ancient than anything that might surface in Troy, Knossos, or Pompeii.

Freud's writing is curiously seductive. The reader is reassured by Freud's frank and frequent admissions that his understanding of the human mind is limited. He openly declares that he is perplexed when confronted by phenomena which psychoanalysis has so far failed to explain. Moreover, he often pre-empts the reader's scepticism with his own. He, too, finds some of his discoveries hard to believe, but has been forced to accept them because of the sheer quantity of evidence. The weight of Freud's hand exerts a ghostly pressure on the shoulder – paternal and consoling – coaxing us to face up to the strange and awesome implications of his work. We are left with the impression of a humble man who, having stumbled upon some difficult truths about the human condition, has reluctantly accepted the burden of taking these truths out into a hostile and unreceptive world. We are urged to pity him and admire him for his courage. Nowhere in his writings is this more apparent than in Lecture 18 of his *Introductory Lectures on Psychoanalysis* (1916–17). In the concluding paragraph of this work Freud claims that, by emphasising the importance of the unconscious in mental life, he had delivered 'the third blow' – after Copernicus and Darwin – to human narcissism.

> In the course of centuries the naive self-love of men has had to submit to two major blows at the hands of science. The first was when they learnt that our earth was not the centre of the universe but only a fragment of a cosmic system of scarcely imaginable vastness. This is associated in our minds with the name of Copernicus, though something similar had already been asserted by Alexandrian science. The second blow fell when biological research destroyed man's supposedly privileged place in creation and proved his descent from the animal kingdom and his ineradicable animal nature. This revaluation has been accomplished in our own days by Darwin, Wallace, and their predecessors, though not without the most violent contemporary opposition. But human megalomania will have suffered its third and most wounding blow from the psychological research of the present time which seeks to prove to the ego that it is not even master of its own house, but must content itself with scanty information of what is going on unconsciously in its mind. We psychoanalysts were not the first and not the only ones to utter this call to introspection; but it seems to be our fate to give it its most forcible expression and to support it with empirical material which affects every individual.

Even among those with a taste for hyperbole, Freud's claim here is arresting in its magnitude. Freud's third blow is not merely the third in a regular sequence, but the third and 'most wounding blow'. According to Freud, the role of the unconscious in mental life was of *greater* significance to humanity than both the heliocentric universe and the theory of evolution.

Many have suggested that Freud's third-blow lecture is definitive proof of his fundamental arrogance and over-inflated sense of self-importance. His modest provisos, cautionary remarks, and sceptical attitude become horribly transparent – nothing more than a weak ploy. Suddenly, they stop concealing the monstrous ego that would lay claim to scientific discoveries of greater significance than Copernicus and Darwin.

This has certainly been the received and most widely endorsed view among members of the scientific establishment.

But what if Freud was right?

5

Darkness rising

In 1902, a Viennese physician called Wilhelm Stekel made a suggestion to Sigmund Freud. He suggested that Freud might convene an informal gathering of local doctors interested in psychoanalysis. Stekel himself was extremely interested in the subject, having only recently been treated by Freud for impotence; however, Stekel was not typical among his medical peers. The Viennese medical establishment were largely indifferent to Freud, and Stekel was one of a very small group of enthusiasts eager to learn more about Freud's methods.

The problem of promoting psychoanalysis had occupied Freud's mind for some time. He had already attempted to raise the profile of psychoanalysis by publishing *The Psychopathology of Everyday Life* (1901) – a book written for the general public which describes how the unconscious can influence trivial errors (such as tongue-slips), accidents, and mistakes. This was, however, a slight work, and over half of the first print run of Freud's magnum opus, *The Interpretation of Dreams*, had yet to be sold. Thus, psychoanalysis was still very much a minority interest.

Freud warmed to Stekel's suggestion, and subsequently sent invitations out to Stekel and three other doctors: Max Kahane, Rudolf Reitler, and Alfred Adler. So it was that on a Wednesday night in the autumn of 1902 four guests arrived at Professor Freud's house – Bergasse 19. The topic they chose to discuss was 'smoking' (an activity which Freud found inordinately pleasurable, and had already decided was a substitute for masturbation). The evening was a great success, and the group decided to carry on meeting regularly. They always convened on Wednesday nights, and in due course their informal gathering became known as the Wednesday Psychological Society.

The Wednesday Psychological Society became something of a local institution. Interested parties were free to come along, some of whom became regulars while others lost interest and dropped out. Nevertheless, slowly but surely, attendance grew. The original members had no real understanding of what they had started. In the fullness of time, the Wednesday Psychological Society would become an international movement and exert a powerful influence on the cultural climate of the twentieth century. What began as five doctors discussing the psychological significance of smoking eventually became something very close to a new religion.

The running order of Freud's Wednesday-night meetings was always the same. One of the group would present a paper, after which coffee and cakes were served. Cigars and cigarettes were also supplied and smoked in large quantities (in spite of Freud's worrying conclusion concerning the real meaning of smoking). Then, fifteen minutes of small-talk was permitted before a group discussion.

Although this agenda sounds very tame, contemporary accounts of the meetings at Bergasse 19 suggest that the atmosphere was far from boring. Participants seemed to enjoy the kind of illicit thrill more commonly associated with the convocation of a cult or underground sect. Even so, there was no stampede to gain cult membership. By 1906 Freud's coterie consisted of only seventeen disciples.

Oddly, as the group enlarged, presentations became more like public confessionals. Many aspirant analysts were inclined to deliver papers on their own sexual problems, early masturbation experiences, or deviant predilections. Presumably this penchant for self-disclosure followed the master's precedent (*The Interpretation of Dreams* is full of candid confessional gems – such as a priceless description of the effect of scrotal boils on Freud's dream life).

In 1907 two Swiss doctors paid a visit: Ludwig Binswanger and Carl Gustav Jung. The following year saw the arrival of the first transatlantic visitor, the Austrian born émigré American Dr A. A. Brill and also Dr Ernest Jones (who would become Freud's most evangelical English disciple). Clearly, word was spreading. So much so that in 1908 the regular members of the Wednesday Psychological Society appointed a secretary and renamed themselves The Viennese Psychoanalytic Society. There was no looking back.

In an extraordinarily short period of time, Freud rose from relative obscurity – the chairman of a local psychology club – to international celebrity. By the 1920s the psychoanalytic movement had branches all over the world, and by the 1930s the image of Sigmund Freud, an elderly man with a beard and an intense, slightly troubled expression, had become synonymous with the mind and its mysteries. Even those with little or no knowledge of academic psychology could claim some familiarity with Freud's ideas and methods.

As the vocabulary and principles of psychoanalysis spread through Europe and North America, so it was that the unconscious became established as a fundamental feature of mental life – relevant to everyday behaviour and experience. A model of mind, in which the unconscious influenced judgement, slips of the tongue, and the content of dreams, gradually insinuated itself in the collective imagination.

Although the concept of the unconscious had been discussed since the time of Leibniz, such discussion had been restricted to the parochial worlds of philosophy, neurology, and psychiatry. Through the vehicle of psychoanalysis, the unconscious 'came out'. Subsequently almost every aspect of cultural life was affected by the 'new psychology'.

The rapid expansion and influence of psychoanalysis was a remarkable phenomenon; however, it was also surprising, given the ambition of Freud's generals. The psychiatrist and historian, J. A. C. Brown, once wrote that Freud 'was surrounded by a group of egocentric primadonnas whose highly ambivalent devotion to the Master was only equalled by their dislike of each other'. But he was probably being generous.

Although the early meetings of the Wednesday Psychological Society were intimate and good humoured, this professional bonhomie was relatively short lived. Petty rivalries soon developed. There were alliances, counter-alliances, squabbles, and frequent episodes of verbal back-stabbing. Moreover, within a few years some members of Freud's inner circle began to covet the old man's chair. In a bizarre re-enactment of Freud's much loved Oedipal situation, Freud's intellectual heirs were soon plotting against him, and after 1910 they began to resemble the murderous primal horde which he was about to describe in *Totem and Taboo*.

The psychoanalytic movement was constantly fracturing and fragmenting. Minor differences of opinion were inflated, leading to bitter arguments and recriminations. Any excuse served to justify breaking away from Freud in order to set up a new school of psychotherapy.

Internecine struggles and ambitious schismatics constantly threatened to raze the edifice of psychoanalysis. Famous devotees like Otto Rank, Sándor Ferenczi, Wilhelm Reich – and even dear old Stekel, all felt the need to break with Freud.

This apostate culture was established early. In 1911 Alfred Adler, one of the original four Wednesday-night meeting invitees, resigned from the Vienna Psychoanalytic Society. His discontent had been fermenting for years. After Freud, Adler was the most senior member of the burgeoning psychoanalytic movement. He had published original and interesting work of his own, and contemporary reports suggest that he was also the most intellectually able of Freud's Viennese satellites. But as far as Adler was concerned, Freud was a colleague, not a mentor or prophet. Subsequently, Freud's pre-eminence rankled.

Adler's approach to mental illness and psychotherapy was very different from Freud's. He was much more interested in context; most notably, social and political. Whereas Freud emphasised internal conflict due to the competing demands of agencies within the mind, Adler emphasised external conflict between the individual and society. The principal motivating force in Adler's theoretical framework was not libido but the desire for power – the need to ascend the social hierarchy. The most fundamental difference between Adler and Freud, however, concerns the status of the unconscious. Although Adler recognised the existence and influence of the unconscious, it does not occupy a central position in his psychology.

When Adler resigned from the Viennese Psychoanalytic Society, his relationship with Freud had reached an all-time low. Thereafter the two men managed to nurse a mutual contempt that lasted until Adler's death twenty-six years later. Records suggest that the news of Adler's death was such a tonic to Freud he could barely stop himself from celebrating. He was particularly pleased to learn that Adler had died in Aberdeen, which he took to be proof of Adler's flagging fortunes and a fitting (albeit belated) punishment.

Nevertheless, in the gallery of psychoanalytic schismatics, pride of place must be awarded to Carl Gustav Jung. Even against the impressive standards set by Freud and Adler, the Jung–Freud débâcle was an outstanding achievement in the field of disastrous personal relationships.

Jung met Freud for the first time in the February of 1907. The two men connected immediately. Their conversation began at one o'clock in the afternoon and, without pause, they spoke continuously until two o'clock the following morning. Afterwards Jung felt that he had been in the presence of greatness – a feeling that was soon reciprocated by Freud.

Jung became Freud's most favoured son. Freud identified him as the 'heir apparent' and even addressed him as the 'Crown Prince'. But their mutual admiration acquired an unhealthy intensity. They began writing to each other like a couple of lovesick adolescents: they missed each other and longed to be with each other – even requesting photographs. This bizarre correspondence reached a kind of apotheosis when Jung finally declared that he had a crush on Freud. It was inevitable that such a feverish, overwrought relationship would end badly.

Like Adler before him, Jung gradually became resentful of Freud's authority. Subsequently he used various theoretical differences as a jemmy to force open a rift between them. The breach was intentionally irreparable, and as it widened their amorous exchanges were replaced by spiteful accusations and personal insults. The late correspondence is brief, formal, and frigid. Indeed, Freud's final letter begins with a stone-cold: 'Dear Doctor'. He felt Jung's betrayal so keenly he couldn't even write his name.

Jung proved to be the most successful of all the post-Freudian heretics. With the exception of Freud himself, no other psychologist developed such a self-contained and comprehensive framework within which to study the mind. And like Freud's psychoanalysis, Jung's *analytical psychology* has implications that extend well beyond psychotherapy. Indeed, Jung's framework resembles a complete metaphysical belief system.

Unlike Adler, Jung was always loyal to Freud's unshakeable conviction in the immense importance of the unconscious. Indeed, it could be argued that within analytical psychology, the unconscious is invested with even greater significance. By probing the very limits of the unconscious, Jung reintroduced a spiritual dimension into western psychology and earned himself a plinth in the pantheon of great thinkers.

For those who found the Freudian unconscious a demeaning maelstrom of primitive animalistic urges, Jung provided an attractive, semi-mystical, alternative. Thus, well before interest in the Freudian unconscious had begun to flag, the Jungian unconscious was already swelling into view. The idea of the unconscious was given a new lease of life, capturing the attention of a fresh generation of psychologists as well the interest of theologians, philosophers, and students of comparative mythology.

Jung's appreciation of the importance of the unconscious began early in his career – long before he had become Freud's 'heir apparent'. His medical degree dissertation was called *On the Psychology and Pathology of So-called Occult Phenomena* (1902). It is a study of a fifteen-year-old medium, Hélène Preiswerk, who was also Jung's cousin. During seances, young Hélène appeared to be possessed by various personalities – mostly deceased relatives and acquaintances. She was extremely convincing. When in a trance state, she would lose her usual Basel dialect and reproduce perfect High German. After studying her, Jung concluded that the voices she produced were not those of her deceased relatives, but 'personifications' arising from her own unconscious.

The idea that spirit guides and possessing spirits were personifications rather than discarnate entities was not a new one. Janet, for example, described a case in 1893 of man possessed by the devil, and subsequently demonstrated a psychological origin. And in 1900 Flournoy published his famous study of the medium Helen Smith. Jung, however, took the idea of personification and made it a key feature of his psychology, where it later re-emerged in the guise of sub-personalities and archetypes.

In 1902 Jung travelled to Paris, where he studied with Janet for a term at the Salpêtrière. Five years later, in a letter to Freud, he would describe Janet as 'a vain old buffer' who only had 'sterile' ideas, but he was probably doing so only to please his new master. Jung later fully acknowledged Janet's influence, and it is again further evidence of Janet's curious talent for invisibility that he is completely overlooked in most accounts of Jung's life and works.

It is interesting that both Freud and Jung were unable to develop their respective systems of psychology before first exploring their own unconscious minds. Freud had undertaken an extensive self-analysis during which dream interpretation proved to be his royal road. Jung also explored his own unconscious, but his methods were far more sensational. For nineteenth-century figures like Coleridge and De Quincey, opium was the key – their way in. For Jung, however, the key was mental illness. After the trauma of breaking with Freud, he experienced a period of extreme mental instability. The mechanisms that hold the unconscious in check broke down, and the unconscious erupted into Jung's life. Instead of resisting this process, he let it take a natural course.

From 1912 Jung felt disorientated and described living 'as if under constant inner pressure'. He suspected that he wasn't well, and that he might be suffering from 'some psychic disturbance'. He began a programme of self-analysis, but the results were inconsequential. Subsequently, he resolved to let the unconscious provide him with the answer to his problems: 'Thus, I consciously submitted myself to the impulses of the unconscious.'

In his autobiographical memoir, *Memories, Dreams, Reflections* (1962), Jung uses volcanic images to evoke his experiences. He describes a 'stream of lava' and the 'heat of its fires' reshaping his life; 'primal stuff' and 'incandescent matter'. The words which he chooses create an impression of extraordinary power being unleashed.

Like Freud, Jung's encounter with the unconscious was highly profitable.

> The years when I was pursuing my inner images were the most important in my life – in them everything essential was decided. It all began then; the later details are only supplements and clarifications of the material that burst forth from the unconscious, and at first swamped me. It was the *prima materia* for a lifetime's work.

Jung kept a record of his dreams and fantasies, which he regarded as communications from the unconscious; however, he also believed that the unconscious could signal its readiness to communicate by engendering unusual urges and impulses in the conscious mind. Jung believed that by surrendering to such impulses he could furnish the unconscious with a much wider expressive repertoire. For example, he indulged in 'childish games' – building a little town out of stones that he found by a lake. The features of this Lilliputian creation could then be interpreted in much the same way as a dream. These experiences were to have a profound effect on Jung's method of practice. He was perhaps the first psychotherapist to suggest that the unconscious might communicate through creative activities; thus, he can be credited with having invented *art therapy*.

Although Jung had surrendered himself to the unconscious, this surrender did not become total until 12 December 1913. On that day Jung was sitting at his desk, contemplating his mental state, when he finally allowed himself to 'let go' of external reality – completely. The experience that followed reads more like an excerpt from a fantasy novel than autobiography:

> Then I let myself drop. Suddenly it was as though the ground literally gave way beneath my feet, and I plunged down into dark depths. I could not fend off a feeling of panic. But then, abruptly, at not too great a depth, I landed on my feet in a soft, sticky mass … Before me was the entrance to a dark cave, in which stood a dwarf with a leathery skin, as if he was mummified. I squeezed past him through the narrow entrance and waded knee deep through icy water to the other end of the cave where, on a projecting rock, I saw a glowing red crystal.

Thus began Jung's visionary journeys into his own unconscious. From this time onwards it seems that the barrier separating imagination and reality collapsed. To describe his experiences, Jung abandons the language of science and adopts a poetic, lyrical vocabulary, much closer in spirit to De Quincey than Freud.

He describes descending into 'empty space' and peering over the edge of a 'cosmic abyss'. He describes travelling to other worlds and speaking to mythical and biblical characters. And he describes his house being invaded by spirits – fugitive projections from his own unconscious.

Among the many exotic personifications that Jung encountered was Philemon, an ancient winged spirit who was particularly vivid: 'At times he seemed to me quite real, as if he were a living personality. I went walking up and down the garden with him, and to me he was what the indians call a guru.'

When Jung finally regained his mental stability he reflected on his bizarre experiences and concluded that, by yielding to the unconscious, he had allowed his mind to heal itself. Different parts of his psyche normally denied expression had been allowed to vent their energies. After fragmenting, his mind had reconstituted itself in a more integrated and stable form. Thus, he believed that he had benefited from a form of 'creative illness'.

This healing process was accelerated through the contemplation of a potent holistic symbol – the mandala (a complex circular design used as a meditation aid by Tibetan Buddhists). Towards the end of the First World War Jung was posted at Château d'Oex, where he established a routine of sketching a mandala in his notebook every morning. He believed that these drawings corresponded closely with his inner state. Thus, he was able to monitor his gradual return to health – the slow assimilation of psychic elements into a new, balanced pattern.

Jung's strange experiences – or, as he would have it, his 'confrontation with the unconscious' – had a profound impact on the conceptual framework of analytical psychology. Like psychoanalysis, analytical psychology is a system that underscores the importance of the unconscious and unconscious processes in mental life; however, compared with Freud's unconscious, Jung's unconscious is an altogether more exotic place. Early on in their correspondence, Freud recognised his junior's predilection for all things mystical. He recognised that his protégé might one day stray too far off the beaten track of scientific respectability, and tactfully urged him not to stay in the 'tropics' for too long. But such advice was wasted on Jung. He had always been – and always would be – fascinated by the paranormal and occult phenomena. Once he had escaped from Freud's influence, it was inevitable that his concept of the unconscious would develop in a more esoteric direction.

Jung proposed three divisions of awareness: *conscious*, *personal unconscious*, and *collective unconscious*. The conscious mind is much the same as Freud described it. The personal unconscious is similar to Freud's unconscious, but differs with respect to some minor, technical details (for example, Jung emphasises the presence of neglected as well as repressed memories). The collective unconscious is Jung's principal modification to the Freudian system. It is the unconscious of the entire species – past and present; however, the collective unconscious also contains racially distinct subdivisions, which can be thought of as a kind of 'folk library' where important cultural symbols are preserved.

The structure of Jung's system is made clearer by use of a simple analogy – the archipelago. Human minds are like individual islands in a chain of islands, and the surface of the sea is like the threshold of consciousness. Below the sea, each island is supported by a unique rock formation. This corresponds with the personal unconscious. Further down, these individual columns of rock join together – they share a common part of the seabed. The level at which local islands join is equivalent to what Jung called the racial unconscious; a storehouse of ancient memories associated with specific ethnic groups (for example Mongolian or Aryan). Descending further, a point is reached where all islands and land masses join. This, the deepest level, corresponds with the collective unconscious.

Jung began to entertain the notion of a collective unconscious while working with patients suffering from schizophrenia. He noticed that, although their personal histories differed, their delusions and hallucinations often shared common themes. Jung also noted certain parallels between the delusions and hallucinations of psychotic patients and the ideas and imagery found in religions and myths. Such observations acquired much greater significance to Jung after his illness, during which he encountered personification of biblical characters and Aryan folk-heroes.

Although the collective unconscious is routinely cited as Jung's most significant contribution to psychology, the thinking behind it was neither unique nor original. The collective unconscious corresponds closely with the old romantic notion of a universal unconscious – an idea that was explored in some detail by a whole generation of poets, philosophers, and artists. Moreover, Freud had flirted with a similar concept for many years before finally committing himself by writing *Totem and Taboo*. In *Memories, Dreams, Reflections* Jung grudgingly acknowledges Freud's priority in a terse footnote: 'Freud speaks of "archaic vestiges"', however, Jung also stresses that Freud's concept of a cultural memory is very different from his own. Whereas Freud's archaic vestiges are like fossilised remnants – dry, dusty, and dead – Jung's archaic vestiges are very much alive. For Jung, the contents of the collective unconscious are a critical influence on psychological maturation and functioning.

Romantic philosophers had already posited the existence of primordial images in the universal unconscious, and suggested that these might emerge in dreams and other altered states. Jung developed this idea within his own theoretical framework and attempted to establish a typology of mythological motifs. He called them *archetypes*. Jung's significant contribution, therefore, was not discovering the collective unconscious, but rather classifying its contents and then suggesting how those contents might influence mental life.

Defining an archetype is not easy. The concept is semi-mystical and subsequently resistant to precise definition. Nevertheless, Jung suggested that archetypes are best construed as psychic 'deposits' – the residua of humanity's repeated encounters with similar experiences, situations, and narratives.

The motion of the sun serves as a particularly good example of a recurring, ubiquitous phenomenon that has clear narrative implications. Every day the sun rises, travels across the sky, and then sinks below the horizon; however, after a period of darkness, it is apparently reborn, bringing light back to the world. Jung suggested that the narrative structure of the solar cycle has acted as a template, facilitating the emergence of countless sun-hero myths. The theme of descent into the underworld, followed by a heroic return, can be found in folklore the world over. This theme emerges not only in obvious literary sources, such as the legends of ancient Greece and Rome, but also in holy books such as the Bible (for example, the death and resurrection of Christ). The sun-hero myth was of course even re-enacted by Freud and Jung, whose periods of self-analysis involved descent into the unconscious followed by a triumphant return.

The sun-hero archetype, although very simple, is a somewhat abstract example. It resembles a principle. Other archetypes, however, are more concrete and exist as personifications – figures that emerge in fantasies, dreams, and visions. When they appear, they tend to arouse feelings of a spiritual or mystical nature.

The spirit-being Philemon (with whom Jung frequently communicated during his illness) is a good example of an archetypal personification. The sage, or guru figure, is a stock image in mythical narratives. Jung believed that such figures share a common origin – the archetype of the *wise old man*. When they appear in dreams or visions, beings like Philemon typically impart knowledge or instruction. They are the means by which the wisdom of the collective unconscious is directed into awareness.

If the manifestations of the wise old man archetype embody superior insight, then the manifestations of the *shadow* archetype demonstrate a very different side of human nature. The shadow organises repressed and unacceptable tendencies – darker and more sinister preoccupations. When personified in dreams, the shadow appears as a primitive protohuman, or an unpleasant figure who arouses apprehension and fear. In myths and folktales, the shadow can manifest in a variety of demonic forms, including the devil. When individuals are overcome with primitive emotions (such as inexplicable rage or the urge to commit violence), Jung suggested this may be attributable to the shadow's influence. It is as if they have become momentarily possessed.

Jung also described two complementary archetypes associated with gender – the *anima* (in men), and the *animus* (in women). The anima organises the feminine characteristics in men, while the animus organises the male characteristics in women. These archetypes could be thought of as 'templates' that shape and guide gender-related behaviour and perception. Thus, Jung suggested that men experience the influence of the anima most in situations requiring emotional sensitivity and intuition, whereas women experience the influence of the animus most in situations requiring rationality and logic.

Perhaps because Jung had more than a mere academic interest in women, he devoted more time to describing the various manifestations of the anima than the animus. He suggested that the anima most commonly finds expression in the figure of an earth-mother – a strong, benign presence symbolising nature and fertility. However, the female character has a more sinister side. This can find expression through a range of other personifications: sirens, seductresses, witches, and femmes fatales. A man's perception of women might be profoundly influenced by the degree to which either good or evil personifications populate his unconscious.

Like Freud, Jung saw the psyche as a battle-ground, with inner conflicts having the potential to result in psychiatric disturbance. Freud limited this conflict to three agencies – ego, superego, and id – each seeking to meet its own particular needs. For Jung, however, the possibilities for internecine struggle are virtually inexhaustible.

The Jungian psyche is not a unitary concept. It is more like a community of sub-personalities, each capable of coming into conflict with any other. Some sub-personalities triumph, becoming dominant, while others are defeated and denied expression. Thus, when a person becomes mentally ill, the mind is quite literally unbalanced.

Jungian psychotherapy seeks to restore equilibrium. The cold logician is encouraged to trust in his or her intuitive side; the extravert learns to appreciate the benefits of meditative self-reflection; and the urban sophisticate discovers the forgotten wonder of seeing the world through the eyes of a hitherto neglected inner child. The different facets of personality are integrated into the characterological equivalent of a mandala; symmetrical, harmonious, and concordant.

Among Jung's extensive writings on the archetypes and the unconscious is a description of a phenomenon he called inflation. Jung observed that individuals who present themselves to the world as leaders – either secular or spiritual – often over-identify with the archetype of the wise old man. When this happens, such individuals come to believe that they are special and that they are party to special knowledge. They become *inflated*.

Although Jung began his career as the most scientific analyst in Freud's circle, he slowly metamorphosed into a holy man. His works became more and more preoccupied with esoterica. He wrote on subjects such as prophetic dreams, the *I Ching*, *The Tibetan Book of the Dead*, and finally suggested that the archetypes might exist (independent of the mind) outside of space and time. In 1958 he suggested that UFOs were omens of an imminent change in the collective mind of humanity.

In later life Jung built himself a fairy-tale house on the shores of Lake Zurich – the location and architecture of which were symbolic, reflecting the contents of his own unconscious. It was a Jungian theme park. Contemporary photographs show a kindly old sage sitting in his garden, reading by a stone block inscribed with alchemical sayings. A modern mystic dressed not in saffron robes, but a light summer suit. It would be difficult to find a better example of inflation.

After the secession of Adler and Jung, many more of Freud's followers decided to break with the Viennese Psychoanalytic Society. Most attempted to establish new schools, which showed varying degrees of fidelity to Freud's original theoretical framework and practice methods. However, the ensuing drift away from orthodoxy was characterised by the repeated emergence of two themes: firstly, the belief that Freud had overemphasised the importance of sexuality; secondly, suspicion that Freud had perhaps overestimated the role of the unconscious.

Apart from Jung – and later, Melanie Klein – fewer and fewer theorists were willing to give the unconscious a privileged position in their revised systems. Attention shifted towards the conscious mind, and particularly the conscious mind in its social context. Influential figures like John Bowlby, Eric Berne, and R. D. Laing, wrote extensively about the role of relationships in the formation of symptoms. Moreover, existential ideas were gradually being absorbed into psychoanalysis. Inevitably, the unconscious seemed less relevant in a universe in which conscious decisions and the exercise of free will were considered the decisive determinants of an individual's fate.

After the 1930s psychoanalysts were less concerned about the contents of the id, and more interested in ego processes. Indeed, so significant was this shift, that a new term, 'ego analysis', came into common usage. The unconscious was not rejected, but rather demoted. Examining the ego, and its relationship to other egos, seemed to offer an exciting way forward – fresh possibilities and the prospect of novel therapeutic advances. Thus, by the 1950s the concept of the unconscious was of crucial significance only to the most loyal devotees of Freudian or Jungian schools.

Yet, as the unconscious was losing the affection of psychoanalysts, it was attracting considerable interest from other quarters – most notably the artistic community. Poets, musicians, painters, playwrights, novelists, and even film-makers, all drew inspiration from the idea of the unconscious, and interestingly, it was the unadulterated, undiluted, Freudian unconscious that captured their attention. The world of forbidden Oedipal wishes, repressed sexuality, and symbolic dreams.

The trespass of Freudian ideas into the wider cultural domain was not welcomed by everyone. For example, the author D. H. Lawrence was sufficiently concerned about the incursion to write two cautionary essays: 'Psychoanalysis and the unconscious' (1921) and 'Fantasia of the unconscious' (1922). Lawrence considered Freud's view of the unconscious too narrow and limiting; however, Lawrence's alternative conceptualisation was somewhat eccentric, and few took heed of his warnings. For many, the strange, disturbing landscape of the Freudian unconscious was too exciting to brook temptation. It exercised a curious fascination.

Predictably, the first works of art showing the unmistakable imprint of Freudian psychology were produced by Viennese artists. The composer Arnold Schoenberg (who lived and worked in Vienna until 1911) actually wrote an opera which takes place in the unconscious. *Erwartung*, which was completed in 1909, is scored for a single soprano and orchestra. The action is spare, but chilling: a woman in a white dress appears at the edge of a moonlit forest and, after following a barely visible path, discovers the body of her lover. He has been murdered. The woman talks to her lover as though he is still alive, but gradually her grief becomes anger and she accuses him of being unfaithful. The woman finally forgives her lover for his past transgressions and bids him goodbye with a parting kiss. *Erwartung* is full of teasing ambiguities, not least of which is the possibility that it is the woman herself who has killed her lover.

This extraordinary work is profoundly Freudian. The libretto is like the fevered confession of a disturbed patient lying on the couch at Bergasse 19. The dark forest suggests the unconscious, wherein – as Freud insisted – opposing emotions such as love and hate can coexist without contradiction, and time and space have little meaning. This is certainly new territory for opera – a nightmare world of supernatural presences, rotting flesh, and violent sexual passion. To describe the Freudian unconscious in musical terms necessitated a radical departure from tradition. Consequently, *Erwartung* was one of Schoenberg's earliest atonal works. The listener is robbed of familiar musical landmarks such as key. Chord sequences fail to make logical progress. This degree of tonal disorientation is sustained until the very last bars, when instead of closing on a satisfying cadence, the music simply dissolves in a flurry of ascending and descending lines.

Another Viennese native who responded to psychoanalysis was the writer Arthur Schnitzler – a man who had much in common with Freud, being not only medically qualified but also very interested in mental illness. Indeed, Schnitzler's first paper, published in 1889, described the hypnotic treatment of several cases of 'hysterical aphonia' (voice loss without an organic cause);

however, unlike Breuer and Freud in the early 1890s, Schnitzler did not find hypnosis very convincing. He believed that some patients were simply 'posing' or 'role acting' and even entertained doubts about the diagnostic validity of hysteria. In the end, he seems to have entertained doubts about the entire therapeutic enterprise. He let his practice shrink, made the acquaintance of several young actresses, and – encouraged by their loose sexual mores – finally decided that his real vocation was writing for the stage.

Schnitzler is now most famous for his novella *Dream Story* (1926), which recently gained a much larger readership when it was filmed as *Eyes Wide Shut* (1999) by the director Stanley Kubrick. *Dream Story* may begin in the real world, but very soon it becomes impossible to establish whether the action is real or imaginary. A wife confesses her secret desire for a young officer and her husband, Fridolin, embarks on a curious nocturnal odyssey that resonates with Freudian symbolism. The text is erotically charged, and a central episode involves Fridolin's illicit entry into a large house, where a masked ball is made considerably more exciting by the presence of statuesque, 'naked beauties'.

By the time Schnitzler was writing *Dream Story* – a relatively late work – Freud's influence on the arts in Europe had extended well beyond Vienna. His ideas were embraced by the French intelligentsia, and most notably by André Breton, the poet and critic who in 1924 launched the surrealist movement with the publication of the first surrealist manifesto, *Le Poisson soluble*.

Breton's goal was to produce works of art that combined elements of conscious and unconscious experience. Thus, he suggested that the world of the unconscious – the world of dreams, fantasy, and visions – should be represented cheek by jowl with the trappings of everyday existence. This super, or absolute, reality Breton described as a 'surreality'. Breton, and poets such as Paul Éluard and Pierre Reverdy, were guided by a creative method or principle that has much in common with Freud's dream work – the process which serves to disguise the latent content of dreams. Their poetry is characterised by odd, seemingly impenetrable juxtapositions, determined not by logic but by unconscious mechanisms.

Another writer associated with both surrealism and psychoanalysis was Georges Bataille. His most famous work, *Story of the Eye* (1928), resembles Schnitzler's *Dream Story* insofar as it is also a sexual odyssey; however, Bataille is far more explicit. Freud's forbidden wishes are fully realised in successive scenes that might best be described as hallucinatory pornography. For example, at one point two men and a woman rape and murder a priest before using one of his eyeballs as a sex toy.

Surrealist art proved considerably more successful than surrealist literature, provoking immediate international interest. Indeed, since the 1920s it has sustained an extraordinary degree of popularity, to the extent that many surrealist paintings have become iconic – being frequently reproduced or parodied in advertisements and the media. This remarkable popularity is easily explained. The bizarre landscapes of Paul Delvaux, René Magritte, and Salvador Dalí are, in fact, curiously familiar. It is as though these artists surrendered to sleep without letting go of their paintbrushes. We recognise the landscape of dreams.

Dalí (who eventually became the high priest of surrealism) actually visited Freud in 1938. Freud was living in London, having fled from Nazi persecution in Vienna. For Dalí, whom Freud later described as having 'fanatical eyes', the visit to Elsworthy Road was not a social call but a pilgrimage. In many respects it was Freud rather than Breton who had written the surrealist manifesto, calling it instead *The Interpretation of Dreams*.

The young Spaniard surreptitiously made a sketch of Freud, which he later turned into a rather enigmatic pen-and-ink drawing. The portrait has a soft, hazy transparency. Freud seems to be fading away. Sadly, neither the sketch nor the drawing was shown to Freud. His family felt that Dalí's picture was rather too accomplished, capturing a presentiment of the old man's imminent death – he was suffering from terminal cancer.

Freud was rather ambivalent about surrealism. He accepted that artists like Dalí were technically very proficient, but he wasn't altogether sure that the unconscious was best captured by odd juxtapositions and fantastic landscapes. Indeed, Freud felt that the presence of the unconscious was more evident in traditional masterpieces – such as the works of Leonardo da Vinci or Michelangelo. The great artists of the Renaissance produced paintings and sculptures that could not be understood without interpretation. The hidden meanings had to be teased out. By attempting to paint the unconscious directly, the surrealists had failed to grasp an essential feature of the unconscious – its mystery.

Freud may have had more sympathy for artists like Jean Arp and Joán Miró. Although they also worked under the banner of surrealism, their approach was somewhat different from the likes of Dalí, Magritte, and Delvaux. Arp and Miró created abstract images that were suggestive, but indefinite. Arp and Miró were not trying to paint the unconscious, but rather to provoke unconscious associations and resonances in the viewer.

Within a few decades psychoanalysis had been transformed from the preoccupation of a group of fringe medical practitioners, to a lingua franca spoken by artists and scientists alike. Yet Freud was to achieve an even greater degree of cultural penetration – much deeper and far wider than anyone had previously imagined. And the reason for this cultural conquest was not a person, or an artistic movement, but a country: America.

In 1909 Freud was invited to America by Professor G. Stanley Hall – a psychologist and the president of Clark University – in order to receive an honorary degree and to deliver a series of talks on the new science of psychoanalysis. Freud had not been championed by any local universities in Austria or Germany – which made Hall's unexpected invitation even more extraordinary. Why should he be honoured by a university in Massachusetts?

The invitation had much to do with Stanley Hall's personal interest in new ideas, and particularly those emanating from Europe. He was an extremely insightful man, who realised long before many of his peers that psychoanalysis was a discipline with considerable promise.

Hall also invited Jung, then Freud's protégé and most zealous advocate; he too was to receive an honorary degree, largely for his work on schizophrenia – his principal area of expert knowledge.

In the end, a party of three analysts set off for America: Freud, Jung, and Sándor Ferenczi.

What Freud called their 'travel adventure' began with a bizarre incident. The three men were having lunch in Bremen, waiting to board ship. Unfortunately, the conversation was dominated by Jung, who spoke incessantly about the results of an archaeological dig taking place in northern Germany. Although Freud was a keen amateur archaeologist, even he was beginning to tire of Jung's persistence. He began to suspect that Jung's obsessive talk of corpses had a hidden meaning; namely, that Jung harboured an unconscious wish to kill him. Freud subsequently fainted.

Later Freud confessed his thoughts to Jung, who responded with utter dismay. Indeed, he claimed to be alarmed at the intensity of Freud's 'fantasies'; however, given that within a few years their friendship would be in tatters, perhaps the old man wasn't so stupid after all. If anyone had an ear for subtext, surely it would be Freud.

In spite of this incident, the following day the trio set off for America in high spirits, determined to entertain themselves by interpreting each other's dreams. After arriving in America the three analysts spent a week in New York before making their way to Clark University.

Jung's letters to his wife Emma give a vivid account of their enthusiastic reception. He met other academics, attended garden parties, and was surprised to find the most unlikely people (two old ladies for example) well informed about psychoanalysis. Jung was particularly impressed by Professor Hall, whom he described as 'refined' and 'distinguished', although he described Hall's wife in less favourable terms – 'plump ... and extremely ugly'. On 13 September Freud and Jung received their degrees at a major awards ceremony. Jung described Freud as being 'in seventh heaven'. At last, he had gained official recognition for his work.

Given these circumstances, Freud should have developed a deep affection for America – the only nation in the world to distinguish him; the only nation in the world to recognise his greatness. In fact, the very opposite happened. Freud had always been somewhat dismissive of America, and after his 'travel adventure' his prejudices became even more entrenched. Indeed, his dislike of America deepened to the point of absurdity. For example, he complained that American food had disturbed his digestive system and that visiting America had caused his handwriting to deteriorate. Moreover, this low opinion of America remained close to contempt throughout his life. Nowhere is this better demonstrated than in his succinct comment on the size of America: 'Yes, America is gigantic, but a gigantic mistake'.

Although Freud's initial anti-American prejudice was entirely irrational, Americans soon gave him good reason to maintain his curmudgeonly attitude. Freud's theories were frequently misrepresented and trivialised in the press, the psychoanalytic vocabulary was misappropriated, and Freud himself was described as some sort of expert on romance. Freud's undisguised disdain had no effect. Even though Freud hated America, America loved Freud. In 1911 both the Psychoanalytic Society of New York and The American Psychoanalytic Association were founded, and in the 1920s psychoanalytic institutes spread throughout the country. Exposure to psychoanalytic ideas became a routine feature of medical training, and the concept of getting to the unconscious root of the problem became axiomatic in the treatment of all mental illness – the goal of all psychiatrists.

The influence of psychoanalysis on cultural life in Europe quickly spread to America. Surrealism was viewed as an important development and many American artists were keen to experiment with this new style; however, psychoanalysis established itself properly in the American imagination not through art but through America's dominant cultural vehicle – the movies.

Perhaps the first person to realise the dramatic potential of psychoanalysis was the film magnate Samuel Goldwyn, who actually tried to get Freud to write him a script. Freud's response to the Hollywood invitation was conspicuously brief: 'I do not intend to see Mr Goldwyn'. It is a measure of Freud's celebrity, that his refusal to co-operate with Hollywood was headline news. On 25 January 1935 the *New York Times* proclaimed: 'Freud rebuffs Goldwyn. Viennese psychoanalyst is not interested in motion picture offer.'

Although Freud was happy to slam the door in Sam Goldwyn's face, Freud's acolyte Karl Abraham wasn't. Subsequently, Abraham collaborated with Goldwyn's studio and the result was a silent film called *The Secret History of a Soul*. The links between film and psychoanalysis were strengthened in the 1930s when many émigré analysts – fleeing Austria and Germany – settled on the west coast. Entering analysis became very fashionable among the studio elite, and Hollywood soon acquired the apposite sobriquet 'couch canyon'. Inevitably, with so many directors, producers, and actors in analysis, films began to show the signs of Freudian influence. Scriptwriters were urged to think very carefully about the unconscious motivation of their characters, and pipe-smoking psychiatrists with earnest expressions began to appear as a stock addition to the dramatis personae of many films.

Among the many Hollywood directors to succumb to Freud's influence was Alfred Hitchcock. Almost of all of his work exhibits a high degree of psychological sophistication; however, in some of his films – such as *Psycho* (1960) and *Spellbound* (1945) – recognition of the mind's darker complexities is made even more explicit. The producer of *Spellbound*, David O. Selznick, was himself in psychoanalysis (as were most of his family). So enthusiastic was he about Freud's ideas that he recruited his own analyst to help him write and vet the script, which, for better or worse, certainly shows. *Spellbound* includes surreal dream sequences and the appearance of an old Viennese doctor – complete with beard and glasses – as a homage to the master. Although *Spellbound* can be criticised for being unsubtle, it nevertheless deserves the accolade of being the first psychoanalytic thriller, and therefore the first cinematic thriller to make significant use of the unconscious as a central concept.

Having influenced art, opera, poetry, the novel, and film, psychoanalysis even managed to inspire a hit broadway musical – *Lady in the Dark*. This extraordinary work was based on a play by Moss Hart, which in turn was based on the writings of his psychoanalyst, Dr Lawrence S. Kubie; lyrics were provided by Ira Gershwin and music by Kurt Weill. As with all psychoanalytically inspired works, dream sequences are its defining feature.

Only in America could the unconscious be represented with so much razzmatazz. Back in old Vienna, Schoenberg's depiction of the unconscious in *Erwartung* had required a cast of one; however, such meagre resources would never satisfy a Broadway audience. *Lady in the Dark* opened in 1941 with a cast of 20 characters, a chorus of 13 singers, 10 dancers, 11 children, and a 20-piece orchestra. Although a Freudian musical sounds like a recipe for disaster, *Lady in the Dark* proved to be a tremendous success. Gershwin and Weill's musical played two seasons on Broadway, toured ten cities, and was then re-engaged on Broadway for a third season. The demand for seats was so great that *Lady in the Dark* established the practice of advanced booking on Broadway.

By the 1950s and 1960s psychoanalysis had scored a remarkable cultural victory. Like a great behemoth, it had stomped over the capitals of Europe and then crossed the Atlantic to conquer the New World. Freud had become a colossus – ranking in stature with the world's greatest thinkers. And the unconscious, the most fundamental feature of psychoanalysis, had become an everyday concept. People were not only discussing the unconscious, they were reading about it in novels, looking at it in paintings, and listening to music – of both serious and popular origin – supposed to evoke the unconscious.

In the first half of the twentieth century Freud's ideas were completely assimilated into American culture. Thus, as America became the dominant world culture – so it was that Freud's ideas were propagated far and wide. Freudian ideas are now totally integrated into the products of American art. But this integration extends well beyond the acceptance of psychotherapy and the appearance of psychoanalysts in the films of Woody Allen. The degree of integration is far more subtle. It works at the level of numerous implicit assumptions made about the mind and human behaviour – sexual desire is profoundly important; all actions can ultimately be explained; secrets can be buried in the mind; and so on.

Although Freud thought America a gigantic mistake, it was a mistake that served him very well.

6

A new vocabulary

Two years before joining Project Ultra and the code breakers of Bletchley Park, Alan Mathison Turing – a rather dishevelled young don from Cambridge – had published an unusual article in a relatively obscure mathematics journal. In this article he had described a device that could read instructions, undertake step-by-step mathematical operations, and store information in a kind of rudimentary memory. It did not escape Turing's notice that he was describing a machine that could ultimately simulate the kind of mental operations performed by human beings when problem solving. Today we would of course immediately recognise Turing's device as a computer.

When Turing wrote his article no one had given the idea of building such a device serious consideration for a hundred years (Charles Babbage being the last). Turing's article did not specify how to build a computer; however, it presented a mathematical model which established the basic elements necessary for a computer to function – should anyone attempt to build one. The pressing needs of war made such an attempt a matter of urgency, and by December 1943 Colossus (widely though incorrectly described as the first electric computer – the Bletchley team were in fact beaten by an American, John V. Atanasoff) had begun to decipher the coded messages of the Nazi war machine with awesome rapidity.

Although few people realised it at the time, Project Ultra would have ramifications that extended beyond the exigencies of war. Indeed, Project Ultra would precipitate a technological and cultural revolution of such outstanding magnitude that no area of human life would be beyond its reach. Even models of the human mind and the unconscious would be utterly transformed.

Turing continued to develop his ideas after the war, but the British failed to exploit Turing's gifts. Moreover, they were slow – or disinclined – to recognise the vast potential of 'electronic brains'. Their transatlantic counterparts, on the other hand, demonstrated a more enlightened view. Funding was made more readily available for men with Turing's vision, and the first general purpose electronic digital computer, ENIAC, was operational at the University of Pennsylvania by February 1946.

Although computer science was destined to become a predominantly American discipline, Turing retained a prophetic role. In 1950 he published an article in the philosophical journal *Mind* called 'Computing machinery and intelligence'. The article begins dramatically: 'I propose to consider the question "Can machines think?"' In attempting to answer this question, Turing managed to prepare the ground for all later work on artificial intelligence.

The suggestion that a machine (composed of insensible components) might generate something like human consciousness had many disconcerting implications. Did the brain function in a similar way? After all, the brain too is composed of insensible components – nerve cells. There were obvious parallels. In the late nineteenth century theorists such as Laycock, Carpenter, and T. H. Huxley had proposed that unconscious mental reflexes might underlie all aspects of human behaviour (including thought); however, this idea, and ideas like it, lost currency after the rise of psychoanalysis. Freud's unconscious seemed far more plausible. In essence, Freud's unconscious was a mind within the mind – a hidden intelligence. It seemed an altogether more reasonable proposition to ascribe intelligent behaviour to an intelligent agency. The appearance of computers, however, reversed the trend. Computers provided a powerful demonstration of how mechanical processes could be organised to produce behaviour that had hitherto been considered exclusively human – such as problem solving and mathematical calculation.

Sadly, Turing did not live to see his ideas on machine intelligence endorsed by the wider scientific community. He died at a tragically young age – committing suicide after his homosexuality resulted in a conviction for gross indecency. His distinguished role in Project Ultra could not be cited in mitigation, as Project Ultra was still classified as top-secret.

For thousands of years contemporary technology has been recruited to help explain the mind. In many respects the history of psychology has been shaped by the selection of metaphors. Wax tablets, clocks, looms, pipes and valves, engines, hydraulic systems, telegraphs, and the telephone exchange have all been employed to represent the mind and mental processes. From ancient times to the middle of the twentieth century the mind was transformed from a marionette (neuron is Greek for 'string') to an electrical maze; however, Turing gave psychology the ultimate metaphor. It was so good, in only a few decades psychologists and neuroscientists were not saying that the brain was *like* a computer, they were saying that the brain *was* a computer.

Brains represent and manipulate information. So, of course, do computers. Thus, when viewed purely as information-processing vehicles, brains and computers appear to be very close cousins. The fact that one is made of organic material and the other inorganic is of less importance than their functional similarity. From the standpoint of an information theorist, they both do the same thing.

Computer technology demanded a new and precise vocabulary. Step-by-step manipulation of information was described as 'serial processing'; the limited amount of information that could enter a system was described as its 'channel capacity'; memory became 'storage capacity' – and so on. This new vocabulary was soon appropriated by psychologists to describe the brain and mental processes.

The sub-discipline of psychology devoted to the study of basic mental processes (such as memory and perception) is called cognitive psychology. Prior to the advent of computer technology, cognitive psychology was somewhat incohesive. Apart from a general commitment to experimental investigations in laboratory settings, early cognitive psychologists did not share a common framework. This situation was immediately rectified by the computer. After the 1950s virtually everyone with an interest in basic mental processes subscribed to the computer model.

One of the publications that helped to launch contemporary cognitive psychology was a 1956 journal article on the size of short-term memory (i.e. the kind of memory used to rehearse an unfamiliar telephone number for the time it takes to make a call). It was written by George A. Miller and tellingly titled: 'The magical number seven, plus or minus two: some limits on our capacity for processing information'. Miller demonstrated that human beings can keep roughly seven units of information (words, names, numbers) in awareness, but rarely more. As such, consciousness can be described as a limited-capacity channel. This simple fact raises some interesting issues.

Given the richness of the human sensory apparatus, the representational capacity of consciousness appears not just limited, but severely limited. At any single moment a human being is registering a considerable amount of sensory information (for example, peripheral objects in the visual field, background noise, and tactile sensations on the body); however, only a tiny fraction of this information ever reaches awareness. How does favoured information get through? Is it selected because of its importance? And, if so, on what basis is that selection made? Almost immediately the information-processing model of mind suggested the operation of unconscious mechanisms.

In the 1940s American experimental psychologists Leo Postman, Jerome Bruner, and Elliot McGinnies had conducted a series of controversial experiments on the influence of personal values on perception. The results were interpreted within the context of psychoanalytic thinking; however, such interpretations were eventually superseded by accounts which relied on the new computer-based model of mind.

A key experiment in their programme of research concerned a phenomenon described as *perceptual defence*. The tachistoscope (a piece of laboratory equipment which can display stimuli for very precise durations) was used to show experimental subjects either emotionally neutral words (e.g. apple, broom, glass) or so called taboo words (e.g. whore, penis, belly). All the words were exposed for incremental durations, and subjects were instructed to identify words as soon as they could. The shortest tachistoscopic exposures were experienced by subjects as nothing more than a flash of light, but as durations increased the words being displayed became easier to identify. Thus, initial exposures were subliminal (i.e. below the limen or threshold of awareness), whereas later exposures were supraliminal (i.e. above the threshold of awareness).

The research team discovered that their subjects required much longer exposure times to recognise taboo words than neutral words. This suggested that some kind of psychological defence was in operation – automatically raising the recognition threshold for taboo words in order to protect the ego from experiencing anxiety.

Although the phenomenon of perceptual defence seemed consistent with psychoanalytic theory, many believed that the results might have a more straightforward explanation. Perhaps subjects were recognising neutral and taboo words at equivalent exposure times, but choosing to delay making a response to taboo words in case they were wrong. Clearly, in a social climate where the word 'belly' possessed improper connotations, the gratuitous expression of more provocative words would have been the cause of considerable embarrassment. The credibility of perceptual defence studies was further undermined by a campaign of more technical criticism, focusing on flaws in the statistical procedures employed to analyse the results. Even so, interest in perceptual defence persisted in several quarters, leading to the design of more sophisticated experiments that were less vulnerable to criticism on both counts (i.e. social and statistical).

In an important series of studies conducted principally by Norman Dixon, of University College London, it was demonstrated that the visual threshold of one eye could be altered by the subliminal presentation of emotional or neutral words to the other. The visual threshold was assessed by gradually increasing the brightness of a spot of light from levels too faint for visual recognition to a level at which subjects could determine its first appearance. Dixon discovered that more intense light was required to ensure visibility of the spot in one eye when subjects were receiving subliminal presentations of words such as whore and cancer (rather than emotionally neutral words) to the other eye. Even though the taboo words did not enter awareness, their presence seemed to be producing a general elevation of the perceptual threshold. In other words, a defensive response. This kind of experiment, which did not require subjects to say taboo words aloud, was beyond the criticism aimed at earlier perceptual defence studies.

How is it that the perceiver can selectively defend against an emotional stimulus unless it has already been identified? Perceptual defence seems to depend on the offices of an unconscious censor, who evaluates all incoming information before making a judgement concerning its suitability for entry into awareness. Perceptual defence is, of course, merely a specific instance of what must be happening all the time if consciousness is a limited-capacity channel.

Many years after conducting the first perceptual defence studies Jerome Bruner colourfully suggested the conscious mind must be equipped with something like a 'Judas eye' – the peephole used by bouncers at speakeasies to distinguish between members (for whom the door would open) and the police (to whom the door would remain closed).

But in a perceptual defence task, who is looking through the Judas eye? Who is opening the door – or at least leaning on it in an attempt to keep it closed?

It seemed unlikely to post-war psychologists that the mind harboured some kind of unconscious secondary personality (or superego), permanently engaged in the task of editing experience before it was deemed suitable for entry into awareness. Subsequently psychologists looked to the new computer-based model of the mind to provide them with an alternative explanation. They reasoned that if the mind functions like a computer, then it very probably processes information in a serial fashion. Perception must involve many tiers of processing, not all of which are accessible to consciousness. Indeed, only the final stages of processing might produce an event in awareness (for example, perception of a word display). Thus a stimulus exposed for a very brief period of time might be processed sufficiently to merit an adjustment in perception threshold, but insufficiently to enter awareness.

Psychologists were still discussing the operation of unconscious mechanisms, but not in the heated language of nineteenth-century Vienna. Instead, they were using the more restrained vocabulary of computer science. The new model did not require unconscious agencies or submerged secondary personalities to explain perceptual defence. All that was needed was a hierarchy of processing stages, progressing from lower to higher levels.

Shortly after the original work on perceptual defence was completed another important study was published in 1951 by R. S. Lazarus and R. A. McCleary. The experiment required subjects to be conditioned – a procedure famously developed by the Russian physiologist Ivan Pavlov. (While investigating digestion in dogs, Pavlov discovered that laboratory animals could be taught to salivate in response to a bell or light if either had previously been paired with food.) Lazarus and McCleary presented meaningless or so called nonsense syllables (always formed by placing a vowel between two consonants, as in QEX or TEG) to subjects, and paired each presentation with an electric shock. The purpose of this procedure was to condition subjects so that they would respond to any further presentations of these syllables with anxiety. Subjects were then shown the shock-associated syllables mixed with a new set of syllables that had not been paired with electric shock. A variety of exposure times were employed, using a tachistoscope. Even when the shock-associated syllables were presented for durations too brief for conscious recognition, subjects still responded by showing increased sweat-gland activity – a reliable indicator of anxiety.

Sweat-gland activity is very easily provoked, and associated with even the slightest changes in emotional state. Although the consequences of excessive sweat-gland activity become obvious in stressful circumstances – as perspiration – most people are unaware of the tiny changes that accompany more subtle emotional shifts (such as when you hear someone saying your name); however, these minute fluctuations are easily detected as changes in skin conductivity if a low current is passed between two electrodes on the skin. The machine that records changes in skin conductance is called the polygraph (although it is better known as the lie detector) and the change in skin conductance itself is known as the skin conductance response (SCR) or the galvanic skin response (GSR). Both terms have equal currency.

The fact that Lazarus and McCleary's subjects produced large SCRs in response to subliminal presentations of shock-associated nonsense syllables was an extraordinary finding. These syllables, being nonsense syllables, had no intrinsic meaning. The only meaning they possessed had been acquired through conditioning. The brain had understood their significance (as things previously linked with pain) and responded by triggering a mild anxiety state; however, recognition and analysis had occurred in the absence of awareness. Lazarus and McCleary described the phenomenon as *subception*, although, like perceptual defence, it is primarily another example of subliminal perception.

Even though phenomena such as perceptual defence and subception suggested that information could be processed outside of awareness, the experimental investigation of the brain's preconscious processing capacity was still regarded by many as a controversial area; nevertheless, the nature of the controversy had changed somewhat. Apart from the perennial presence of those who seemed to object to anything relating to the unconscious as a matter of principle, the existence of preconscious processing was not in doubt. The real issue concerned the degree to which information could be processed in the absence of awareness.

If the brain does work like a computer, then incoming information will be analysed in a step-by-step fashion, where each step is associated with a more complex analysis. For example, the first stage might involve a very basic registration of presence (Is something there or not?); the second stage might perform a structural analysis (What does it look like?); a third stage might seek to establish emotional significance (Is it a good thing or a bad thing?); and finally, a fourth stage might execute a much more sophisticated semantic analysis (What does it mean exactly?). To what extent then, are subliminal stimuli processed? Some commentators have chosen to ask the same question using less technical language: 'Is the unconscious smart or dumb?'

Through the observation of patients, Freud and his followers had come to the conclusion that the unconscious was very smart indeed; however, hardly any experimental work had been undertaken to prove this. With the arrival of instruments like the polygraph and the tachistoscope, psychologists were suddenly in a position to begin putting the unconscious through its paces. For advocates of a smart unconscious the early signs were promising – phenomena such as perceptual defence and subception suggested that stimuli could be analysed for meaning in the absence of awareness. But not everyone was convinced. Thus, the experimental agenda in subsequent years was very much concerned with exploring the degree to which preconscious processing could be described as superficial or complex.

In 1953 Colin Cherry published an article with the intriguing title: 'Some experiments on the recognition of speech with one and with two ears'. He had become interested in understanding a phenomenon called the cocktail party problem, which concerns the brain's ability to follow a conversation against the general hubbub of a party. How is it done? What features of a single human voice, among many others, does the brain lock on to? To explore the cocktail party problem, Cherry introduced a new experimental procedure called the *dichotic listening task*. Experimental subjects were given a set of headphones and asked to follow a message which was played through one channel while an entirely different message was being played through the other channel. Thus, the left and right ears were receiving separate streams of information. Cherry discovered that the ability to follow a message in one channel (while ignoring the message in the other) was largely dependent on 'physical' characteristics; for example, the gender of the voice or its location. When Cherry presented recordings of two different messages but spoken by the same voice – thus eliminating any superficial differences – listeners found it difficult to follow a single channel. The different messages got mixed up.

Cherry also conducted experiments in which one message was shadowed (i.e. immediately repeated back aloud) while a second message was played in the other ear. When subjects did this, very little information from the unattended message 'got through'. For example, subjects didn't notice when the unattended channel contained passages spoken in a foreign language. Nevertheless, so-called physical changes (such as the introduction of a pure tone) were always detected. Cherry concluded that the unattended message received little or no processing.

Cherry's work is important for several reasons. Firstly, he more or less started systematic research into the subject of focused attention; secondly, by devising the dichotic listening task, he equipped experimental psychologists with a simple method of controlling the presentation of auditory information to subjects. Both the study of attention and the dichotic listening task would play an important role in subsequent research into the unconscious and pre-conscious processing.

The study of attention came to the forefront of academic psychology with the publication of Donald Broadbent's *Perception and Communication* (1958). This was a landmark work because in it Broadbent describes a model of attention that owes much to electronic communications theory and computer technology. It assumes that the brain is best construed as an organ that processes sensory information through limited-capacity input channels. Moreover, Broadbent chose to illustrate different stages of processing with the aid of flow charts (a device already favoured by computer programmers).

Broadbent's model was originally devised to explain how the brain copes with all of the information it receives through the senses. He suggested that the brain possesses a central information-processing channel of very limited capacity, which corresponds roughly with awareness. The central channel can only select one sensory input channel at a time, but can switch between channels at the rate of about twice every second. When we 'pay attention' to something, we are selecting a specific input channel for more detailed processing in the central channel. Information that is still coming in through unattended channels occupies a short term memory store for a few seconds only. Then it is lost. Broadbent called the mechanism that selects incoming information for more detailed processing the *filter*.

After Broadbent, other psychologists proposed similar theories, often including a selection mechanism resembling Broadbent's filter. Subsequently, the term 'filter theory' was routinely employed as a generic term.

All filter theories posit a bottleneck in the information-processing apparatus. The vast amount of incoming sensory information encounters a limited-capacity channel, which necessitates the selection of some stimuli for representation in awareness at the expense of others. If this were not the case, then human beings could simply attend to everything. The concept of 'paying attention' automatically implies the existence of a central processing mechanism which regulates the passage of information through the bottleneck.

Broadbent supposed that incoming information was selected for further processing at a very early stage (i.e. immediately after a superficial physical or structural analysis); therefore, Broadbent's theory is described as an *early selection* theory. A *late selection* theory would entail a more thorough analysis (involving perhaps analysis of stimuli for meaning) before the selection was made. In either case, selection would eventually result in the admission of incoming information into awareness. Broadbent's early selection theory provided a good framework for understanding the results that Cherry had reported after conducting his dichotic listening studies. While shadowing a message in one ear, experimental subjects could only identify superficial features of the message playing in the other ear (e.g. the gender of the speaker). Broadbent's theory suggested that unless information entered awareness, it could only ever receive a superficial analysis. As far as Broadbent was concerned, then, the unconscious was not smart, but fairly dumb.

Within a year, however, the legitimacy of Broadbent's theory was called into question. Employing Cherry's dichotic listening task again, it was discovered that some words in the unattended message *are* processed for meaning. For example, if the name of an experimental subject is placed in the unattended message it is likely to be noticed.

Most people have had a direct experience of this phenomenon. If someone mentions your name at a party it tends to rise above the background noise, even if spoken quite softly. The background noise is the equivalent of an unattended message in a dichotic listening task. You are not conscious of monitoring the background noise, but to some extent, you must be – or you wouldn't hear your name.

The finding that some words in the unattended message are processed for meaning suggested that the brain's filter operates a late selection, rather than an early selection principle. All of the information arriving through the unattended channel receives a semantic analysis, to the extent that personal names can be identified, tagged as important, and subsequently admitted into awareness.

This discovery raised other issues. For example, if all incoming sensory information was given a thorough analysis, what happened to information that was analysed, but still not selected for entry into awareness? Could such information – now invested with meaning – influence the processing system in a more general sense?

It is interesting that, in spite of all the work conducted on perceptual defence, subception, and filter theories in the 1940s and 1950s, interest in preconscious processing diminished dramatically in the next decade. There are many reasons for this. Unfortunately, preconscious processing was still linked in the minds of many psychologists with the old psychoanalytic and romantic unconscious. To suggest that parts of the mind enjoyed a certain amount of independence smacked of hypnotism, spiritualism, and exotic sub-personalities. The nineteenth century in fact. In a post-war world keen to embrace a bright new future, many were disinclined to investigate a phenomenon that seemed tarnished by a somewhat disreputable past. Cognitive psychology had only just been provided with a powerful new metaphor and vocabulary – it wanted to distance itself from any hocus-pocus. Consequently a number of cognitive psychologists sought to enhance the scientific standing of their discipline by attacking any work purporting to examine the preconscious processing capacity of the brain.

Such criticism certainly had prima facie validity. Many argued that the unconscious plays a relatively minor role in the context of day-to-day mental functioning; therefore, logically, cognitive psychologists might be more profitably engaged in the study of more substantial phenomena (e.g. memory and reasoning). Moreover, they argued that the experimental procedures employed to investigate the unconscious militate against common sense. As a general principle, we are not affected by things that we don't experience. If an unpleasant car accident occurs five miles away from our current location we do not feel any distress. Why should we be affected by events and information occurring beyond awareness? The idea that psychological and physical reactions might be provoked by exposing subjects to stimuli at such low intensities that they were effectively 'not there' seemed wholly implausible.

In a short space of time the official information-processing model of the brain rejected the unconscious altogether. It was assumed that unattended stimuli and latent memories were not capable of affecting conscious mental operations, and that the results of experimental studies suggesting otherwise could be dismissed on account of their poor design. (In other words, positive results had been mistakenly attributed to preconscious processing, when in fact they could be accounted for by other factors.)

Finally, in a decade dominated by liberal ideas, the concept of conducting experiments in which the participants are necessarily deceived was anathema. Clearly, in studies designed to investigate preconscious processing, experimental subjects cannot give informed consent with respect to their involvement. This would entail discussion of specific subliminal stimuli, thus undermining the subliminal nature of the experiment. Non-consensual experimentation seemed even more offensive after the so called 'subliminal scandal' of the late fifties. Subliminal messages were employed by an advertising company to increase food and drink consumption in American cinemas. (This affair will be examined more closely in Chapter 9.) The subsequent furore helped to bury research into preconscious processing for the next decade – or so it appeared.

In 1971 Norman Dixon published *Subliminal Perception: The Nature of a Controversy*. In this comprehensive work, Dixon carefully examined the evidence for subliminal perception; it was later much expanded and reissued in 1981 as *Preconscious Processing*. This encyclopaedic review is not only an outstanding work of scholarship but also a kind of ark. It ensured the survival of important work that might otherwise have sunk into obscurity. This is particularly true of many studies conducted during the period when research into preconscious processing had become deeply unfashionable.

Fortunately, Norman Dixon was an individual who displayed a healthy disregard for academic fashions and partisan squabbles. This may have had something to do with his unusual background. Before opting to study psychology in 1950, Dixon had experienced ten years of service in the Royal Engineers, nine of which were as a regular officer in the bomb-disposal team. During this time, he was severely wounded (requiring the amputation of an arm). Even so, Dixon possessed an extraordinary, sanguine temperament, and cheerily maintained that he had sustained the wound through his own incompetence. He later demonstrated a gift for rescuing something positive from personal misfortune by writing an original and entertaining book titled *On the Psychology of Military Incompetence* (1976) – a work highly regarded by the military. In 1974 Dixon was awarded the University of London Carpenter Medal for work of exceptional distinction in experimental psychology.

In the foreword to *On the Psychology of Military Incompetence* Brigadier Shelford Bidwell describes Dixon as 'a bold man', with a character 'moulded in a corps where intellect habitually meets danger'. When Dixon chose to champion preconscious processing, he chose an area where, again, intellectual demands habitually met danger. *Preconscious Processing* is a highly erudite work which expertly guides the reader across an academic minefield. Just the kind of accomplishment that one would expect from a man all too familiar with the consequences of making a wrong move in a hostile environment.

Dixon's conclusions are solidly in favour of the reality of preconscious pro-cessing. He suggests that, typically, preconscious processing involves a sophisticated analysis of subliminally presented information. Moreover, sub-liminally presented stimuli – presented in any sensory modality – may have a marked effect on the conscious mind. Perception, motivation, thought, and emotional state, can all be influenced by subliminal stimuli, as can the func-tioning of memory, the content of dreams, and the progress of a psychiatric illness. Dixon also suggested that experimental evidence favoured the some-what counterintuitive view that subliminal presentations of decreasing inten-sity are accompanied by larger effects. In other words, the more brief or faint the subliminal exposure, the more significant its consequences.

Dixon supports his conclusions by examining a vast body of experimental research, some of it little known and dating back to the nineteenth century. Moreover, Dixon draws from several and diverse sources. Unlike many parti-san theorists, he is comfortable with both psychoanalytic and cognitive frameworks. Consequently processes that operate outside awareness are viewed with welcome neutrality. The terms unconscious and preconscious processing had acquired very different connotations: the former associated with the hot, teeming world of the id and the latter associated with the cool, mechanical world of the computer. However, in many respects, the two terms are merely different ways of describing the same stratum of mental life. Thus, in Dixon's work, Freudian concepts are discussed in chapters that also include flow charts showing the serial processing of information.

Dixon's theoretical equanimity is characteristic of a trend that developed in the 1970s, which involved the re-examination of Freudian ideas through the lens of contemporary cognitive psychology. Guided by the computer metaphor, theorists were talking about filtering and selectivity rather than censorship; executive processes instead of the ego; working memory instead of consciousness. Suddenly, there were routines, sub-routines, programmes, and software, rather than psychic structures and agencies of the mind. The new vocabulary helped legitimise discussion of certain Freudian ideas that only a few decades earlier had been rejected as absurd and unscientific. Thus, the psychoanalytic edifice was renovated. Its fussy nineteenth-century orna-mentation was removed, giving it a sparer appearance, more pleasing to modern preferences. Once this process was complete many theorists were able to consider the concept of the unconscious again with a clear scientific conscience, and many subsequently concluded that something vital had been missing from mainstream models of the human mind for many years.

Although mainstream psychology had taken a dim view of the unconscious during the 1960s, cognitive psychology had in fact produced models of the mind which assumed preconscious processing stages without ever making these explicit. Certain boxes in their flow diagrams were not shown – but must necessarily precede the stages of processing described in other boxes. The renovation of psychoanalytic ideas had made discussion of the unconscious acceptable again. It was now possible to believe in the unconscious without loss of credibility. Consequently the unconscious – as described using the new vocabulary – was reintroduced into mainstream academic textbooks. It was soon permissible to say things that Freud himself would have agreed with – but without being in any sense a Freudian. By the early 1980s textbooks had begun to suggest that conscious phenomena had perhaps been overemphasised as most mental operations must necessarily take place outside of awareness. In an extraordinary U-turn, preconscious processing was not only uncontroversial but a fundamental feature of virtually all information-processing models of the mind.

The old distinction between conscious and unconscious domains was subsequently reformulated in terms of *controlled* and *automatic* processing. The former is initiated by an act of will, is maintained through sustained effort, and progresses in a serial fashion. The latter is likely to be triggered by environmental demands or stimuli, is maintained without effort, and can involve procedures that progress in parallel. Calculating an arithmetic sum is a typical example of controlled processing whereas driving a car on a familiar route is a typical example of automatic processing. The first task requires conscious attention while the second requires virtually none. A skilled driver on a familiar route can concentrate on a conversation or listen to the radio while all the physical movements necessary for driving are simultaneously co-ordinated in response to continuously changing environmental demands, without conscious effort.

The concept of automatic, unconscious processes operating in the brain had appeared erratically in philosophy, psychology, and neurology since Leibniz; however, the computer metaphor allowed such processes to be understood in a more satisfying way. They could be construed as a program. Once initiated, automatic processes tend to run a predetermined course unless interrupted. The fact that such automatic programs exist in the brain was supported by work on what came to be known as *control* or *capture errors*. When distracted or preoccupied, people often do things that they don't mean to. For example, they might absent-mindedly open a cupboard door, switch on a light, or put on a second pair of glasses. When unintentional actions take place they tend to be selected from a repertoire of very common action patterns. It is as though, while consciousness is otherwise engaged, a triggering stimulus is analysed outside of awareness and accidentally 'switches on' a program that can automatically orchestrate a routine set of connected behaviours. The program may run to completion, or, alternatively, the program might be interrupted if the person becomes aware that he or she is functioning on 'autopilot'.

Some automatic programs are clearly innate – for example the program in the brain responsible for keeping the heart beating; however, automatic programs of the kind so far discussed – the kind activated while opening cupboards or driving cars – are clearly learned. First, the elements of the skill are co-ordinated through controlled processes; however, after much rehearsal, performing the skill requires less and less mental effort. Eventually the skill is executed more or less automatically. Indeed, what we call proficiency is really the degree to which a skill has become automated. It is interesting to note that when a highly complex skill has become automated, individuals lose the ability to explain how it is performed. Professional athletes and musicians, for example, are notoriously bad at analysing their own abilities. Moreover, they tend to shy away from self-analysis because self-awareness during the operation of automated skills seems to cause problems – mistakes are made more frequently. Once a program has been installed in the unconscious it is best left alone.

The tendency for skills to become automated seems to reflect a general principle that guides the operation of the brain. As soon as a skill is learned, it is automated as swiftly as possible. This makes a great deal of sense, because consciousness – although limited and slow – is absolutely necessary when it comes to dealing with new situations. Controlled processes are essential for planning and the acquisition of new skills. Consequently, achieving maximum availability of controlled processing resources is extremely useful.

From an evolutionary perspective this arrangement has obvious advantages. An organism that deals with familiar situations quickly and efficiently, but is always ready to adapt to new challenges, is surely optimising its chances of survival.

Once again, the transfer of skills from conscious to unconscious domains was reflected in a procedure familiar to computer technicians – *knowledge compilation*. This is the procedure that changes the format in which data is represented.

Automated behaviour suggests blind, mechanical processes – implying that the unconscious is not very intelligent; however, blind, mechanical processes, when operating quickly and in parallel, are capable of producing remarkable results. A particularly good example is the so-called *time-gap experience*. After a long journey it is not uncommon for motorists (and particularly long-distance lorry drivers) to arrive at a destination and have no conscious recollection of how they got there. When the conscious mind is absent without leave, the unconscious can take the wheel and demonstrate extraordinary levels of competence.

As the notion of preconscious processing became more widely accepted, it continued to gain support from various sources. In the same way that unintended actions associated with absent-mindedness could be viewed as examples of preconscious processing, so it was that other everyday phenomena succumbed to the same analysis.

Human beings are constantly making use of unconscious knowledge. With respect to language, the rules that determine sentence construction are applied in a completely intuitive way. Most people would accept that 'the big blue car' is better English than 'the blue big car'; however, apart from the odd language specialist, few would be able to justify their opinion. The syntactical and semantic rules that govern word selection operate from below the awareness threshold and are largely inaccessible. Experimental investigations began to demonstrate that many aspects of human behaviour are influenced by inaccessible rules and heuristics.

In 1977, psychologists Richard Nisbett and Timothy Wilson, published a now celebrated article titled 'Telling more than we can know: verbal reports on mental processes'. Experimental subjects were asked to assess the quality of four *identical* pairs of stockings. Subjects made choices, and justified their choices with reference to the quality of the material. In fact, it transpired that stockings judged to be of better quality were simply those presented on the right-hand side. Of course, being identical, there were no real differences between the stockings. The reason generated by subjects to justify their choice was merely a post-hoc rationalisation.

More sophisticated experiments revealed that people readily acquire knowledge in the form of rules, but are wholly unaware that such learning has taken place. In a typical experiment of this kind, subjects are asked to watch a visual display on a computer screen. The display changes and develops according to a highly complex set of rules – too complex to be worked out during the course of the experiment. After watching such displays, subjects are then asked to make predictions (e.g. Where will the white circle appear next?). Studies of this kind have found that subjects perform remarkably well, making very accurate predictions; however, when asked to explain their reasoning, most usually confess that they were just guessing. Most have absolutely no idea why they were choosing to make one prediction rather than another. Clearly, investigations of this kind demonstrate that complex rules can be learned and subsequently employed to guide judgements without ever rising into awareness.

One of the most invariant characteristics of human beings is their tendency to form favourable preferences based on familiarity. The technical term for this phenomenon is the *mere exposure effect*. If shown pictures of geometric shapes, buildings, or people – anything in fact – experimental subjects have a marked inclination to judge pictures that have been shown at least once before more pleasant than comparable pictures that haven't. In the laboratory, at least, familiarity breeds not contempt but affection.

In 1980 an important study was published by William Kunst-Wilson and Robert Zajonc. Experimental subjects were exposed to subliminal figures (polygons), which were then presented supraliminally to the same subjects – but mixed up with new figures of similar design. Although experimental subjects did not recognise any of the supraliminal presentations, they demonstrated a significant preference for those that had been previously shown to them subliminally. This study prompted many other investigations that produced the same or similar results – demonstrating that familiarity-based preferences are substantially influenced by unconscious memories.

Another everyday phenomenon thought to be influenced by preconscious processing was social perception. For human beings, first impressions appear to be very important. They are formed very quickly and, more often than not, without much consideration. Perhaps the most extreme instance of a swift and favourable social judgement of this kind is 'love at first sight'.

In an attempt to understand the mysterious processes that govern social perception, many psychologists began to look to the unconscious for answers. It was suggested that first impressions might be mediated by unconscious information. Subsequently tasks were devised in which subliminal words (relating to character traits) were presented to experimental subjects just before they were required to form an opinion about persons for whom they were given no other information. Again, a strong relationship was found between the tone of subliminal presentations and subsequent judgements.

In addition to the growing body of empirical support for the unconscious, a major theoretical advance helped to consolidate the unconscious as a legitimate concept in contemporary neuroscience. This new advance was an extension of information-processing theory known as *connectionism* or *parallel distributed processing* (PDP). This approach was something of a hybrid, combining the computer model with extant knowledge of biological mechanisms in the brain.

The PDP model postulates the existence of *independent processing units*, each capable of accomplishing a specific (but very basic) task. When active, each unit excites or inhibits other units in an extended network (which resembles the network of neurones in the brain). Eventually, all of this frantic activity is resolved in a stable pattern of mutual influence, which represents the information being processed.

The PDP model has a number of features which were of interest to those studying the unconscious. Firstly, the PDP system does not contain a single central processor – the equivalent of a place where consciousness is synthesised. The PDP analogue of consciousness emerges from an interacting system of insensible processing units. Secondly, the level of activation in each processing unit can vary. Thus, a subgroup of processing units might become gently active, affect the whole system, but not achieve conscious representation. In other words, PDP allows behaviour to be influenced by subliminal events. Finally, the PDP model replaced serial processing with parallel processing. Thus, a vast number of processing units are in conversation at any particular moment in time. The number of active units and the speed with which they deal with information necessarily exceeds the capacity of consciousness.

By the end of the 1980s a considerable amount of research had been undertaken, exploring the unconscious from within the new cognitive framework. Studies had been conducted on unconscious perception, unconscious learning, unconscious memory, and the acquisition of automated skills. The computer metaphor had been extremely fruitful. It had inspired a new generation of experimental psychologists to re-examine a concept that had lost scientific credibility in the custody of psychoanalysis, and had been neglected altogether by behaviourists. A large body of experimental evidence had been gathered, demonstrating the importance of preconscious processing, which in turn attracted the attention of other respectable disciplines; for example, neurologists and assorted brain scientists were also taking a more keen interest in the new unconscious. In addition, theoretical advances like the PDP model were entirely consistent with the view that information processing in the brain is a largely unconscious activity.

In 1987 the psychologist John Kihlstrom published a brief article in the journal *Science*. It was titled 'The cognitive unconscious'. In this article Kihlstrom reviewed numerous lines of research (including some described above), which, taken together, demonstrated the emergence of a new model, a new empirical ethos, and a new unconscious.

A key passage in this article concerns the scientific legitimacy of automated processes. Kihlstrom accepts the evidence that in specific circumstances (as demonstrated particularly in laboratory studies), meanings and implications can be understood in the absence of awareness. Consequently individuals can make a judgement (for example, whether they like someone), and then act on that judgement, without any knowledge of how it was reached. Kihlstrom stresses that this does not mean that genuine cognitive activity has not taken place. Rather, it means that the process of judgement or inference has become automated, and hence unavailable for introspection.

Although the unconscious might seem dumb, it is, after all, very smart. In many respects, it is re-formatted intelligence. Kihlstrom's final conclusion is dramatic. Although constrained by the formal requirements of journal writing, he is unable to conceal a flare of genuine excitement.

> One thing is now clear: consciousness is not to be identified with any particular perceptual-cognitive functions such as discriminative response to stimulation, perception, memory, or the higher mental processes involved in judgement or problem solving. All of these functions can take place outside of phenomenal awareness. Rather, consciousness is an experiential quality that may accompany any of these functions. The fact of conscious awareness may have particular consequences for psychological function – it seems necessary for voluntary control, for example, as well as for communicating one's mental states to others. But is not necessary for complex functioning.

A new model. A new ethos. A new unconscious.

Yet, if you compare this excerpt with Freud's famous 'third-blow' lecture, it is difficult not to conclude that, fundamentally, they are both saying the same thing: the unconscious has been very much underrated.

When Freud first arrived in London in 1938 he was accommodated very briefly in Little Venice. Although he had no way of knowing it at the time, the house he was staying in would one day display a commemorative plaque bearing his name and the name of a man who had been born in the same house twenty-six years earlier. Eventually, it was decided that Freud's short sojourn in Little Venice was too insignificant to merit commemoration, and another plaque was mounted exclusively devoted to the birth and short life of Alan Turing.

The old plaque was a rare (and perhaps the only) instance of Freud and Turing being honoured together in a public place. This is hardly surprising as – at first sight – they seem to occupy entirely different universes. The name Freud conjures images of nineteenth-century Vienna, hysteria, and the landscape of dreams, whereas Turing conjures images of Bletchley Park, mathematics, and the bright new world of information technology. However, in their very different ways, these two men did much the same thing. They both provided the twentieth century with a model of mind. Both of these models – in one way or another – dominated psychology for the entire century.

Ostensibly, the psychoanalytic model and the computer-inspired cognitive model are poles apart. Nevertheless, by the end of the 1980s, with respect to the unconscious, the traditions of psychoanalysis and cognitive psychology had demonstrated a curious convergence. Although they had started from very different positions, their destinations were the same. On the big issues concerning the nature of mental life, they agreed. Yes, the importance of consciousness was overrated, and yes, the unconscious was far from stupid.

The hidden intelligence was alive and well, but now simply residing under a different name.

7

The unconscious brain

In the 1930s and 1940s the Canadian neurosurgeon Wilder Penfield conducted an extraordinary series of experiments that dramatically demonstrated the intimate relationship between brain and mind. His work represented a troubling challenge to the advocates of Cartesian dualism. The 1.4 kg of gelatinous matter that comprises the human brain was shown to be the organ of consciousness – the physical reality behind the phenomenal world.

Penfield was originally concerned with locating the foci of epileptic seizures in the brain. Once this was achieved, diseased tissue could then be removed by surgery. The problem was identification. How could these suspect areas be located?

Individuals with epilepsy characteristically report unusual sensory experiences and impressions before a seizure occurs. These warning signs – which differ from person to person – are known as auras. Penfield reasoned that if the cerebral cortex of the brain was exposed and electrically stimulated, this might provoke an aura, thus indicating the location of the diseased tissue site. The cortex covers the surface of the brain and mediates higher mental functions such as conscious perception and thinking.

In order to undertake this procedure Penfield anaesthetised his patients and cut around the circumference of their skulls – he was thus able to expose the cortex by raising the crown of the head like a lid. Because the brain does not register pain directly (it can only register pain via specific nerve pathways from the rest of the body), Penfield was able to place electrodes on the cortex without causing his patients any discomfort – that is, apart from the obvious psychological discomfort that must necessarily accompany having a surgeon play with one's brain.

The surgical removal of abnormal tissue proved to be an effective means of ameliorating epileptic seizures; however, Penfield's permanent entry in neuroscience textbooks was secured not on account of his clinical successes but because of the responses he elicited from patients while stimulating their brains with an electric probe.

When stimulated, specific areas of the brain produced specific responses. These varied from involuntary movements to a wide range of sensory experiences – tingling, feeling hot or cold, seeing flashes of light or abstract patterns, hearing clicks and buzzing. Some patients reported experiences of a far more complex nature, in the form of visual and auditory hallucinations. One, for example, thought he was being chased by armed robbers (who came at him from behind and to the left). Another heard a Beethoven symphony so clearly that the patient accused Penfield of concealing a radio.

Perhaps the most interesting responses that Penfield elicited were those associated with electrical stimulation of the temporal lobes. These structures form much of the lower side of each half of the brain and are positioned just above the ears. Patients reported a range of odd experiences such as *déjà vu*, or its exact opposite, *jamais vu*. They also experienced different mood states – fear, foreboding, and euphoria; however, the most extraordinary reports that Penfield elicited were of so-called dreams. These took the form of vivid hallucinatory experiences in which the patient became the protagonist in a short narrative. These experiences were never very elaborate, but involved the appearance of characters ('There are a lot of people'), awareness of different environments ('… an office somewhere'), and sounds ('I heard boom, boom, boom'). These experiences were described as dreams, and sometimes possessed the unique, inexpressible quality of dreams – making it hard for patients to offer a precise account of what was happening to them.

Penfield seemed to have worked a miracle formerly necessitating opium, hypnosis, or insanity – he had liberated the contents of the unconscious. Moreover, unlike his romantic and psychoanalytic predecessors, Penfield seemed to have discovered where the stuff of dreams collects – in the deep crevices of the temporal lobe.

Was this, then, the physical location of the unconscious? Had Penfield shocked the unconscious out of its hiding place? And, more importantly, had Penfield finally demolished Cartesian dualism – reducing the mind (in all its complexity), to an electrochemical jelly?

Clearly, the tacit agenda of Penfield's work was reductionist in nature. He had assumed that all mental experiences must be attributable to electrical and chemical activity in the brain. Over many years, the evidence he had gathered together was entirely consistent with this view; however, in later life he expressed a curious ambivalence concerning the value of his work, and cautioned others against simplistic interpretations. Moreover, the year before he died, he insisted that neuronal activity in the brain could never account for such human qualities as freewill, and argued for a distinction between brain and mind. Although there were some minor differences between his position and that of Descartes, such differences were only in the detail. Bizarrely, Wilder Penfield died a card-carrying Cartesian dualist.

The idea of a conscious mind exercising freewill is so powerful, and so central to the concept of being human, that even a reductionist like Wilder Penfield struggled to preserve its sanctity. It seemed – and still does seem – highly improbable that something as fickle, fleet, and spontaneous as thought can be the result of purely physical processes in the brain. Newtonian physics is mercilessly deterministic. So how can freewill be reconciled with the electrochemical foundations of behaviour – which must necessarily operate within the limits of unyielding physical laws? In the absence of a non-physical consciousness, each current brain state must be determined by its preceding brain state. Yet we all feel that we have choices.

Irrespective of the philosophical implications of Penfield's work, the surgical and exploratory procedures that Penfield developed represented a considerable technical advance. He had demonstrated that it was possible to manipulate the brain of a conscious patient, and thereafter any surgical procedure that involved exposing the brain was also recognised as a potential opportunity to extend and develop Penfield's work. Yet it was an opportunity that remained largely unexploited until the late 1950s. It was then that an astonishing research programme was initiated, one that would have profound implications for neuroscience in general and constitute a major chapter in the history of the unconscious.

In 1958 Bertram Feinstein, a surgeon working in San Francisco, was developing a procedure devised to treat the tremors, tics, and spasms that resulted from certain types of brain disease. A small hole was made in the skull through which electrodes could be inserted. A relatively strong current could then be passed through the electrodes, thus destroying the abnormal cells which Feinstein believed were causing the involuntary movements.

Feinstein was a friend of Benjamin Libet, a physiologist based at the University of California School of Medicine. Libet was already studying brain chemistry, and was very interested in Feinstein's work. Moreover, he was keen to experiment with the living brain, replicating and perhaps building on Penfield's studies.

The surgical procedure that Feinstein employed was more refined than Penfield's, which meant that Libet had restricted access to the cortex. To conduct his operations, Penfield had removed the top of the skull completely. Feinstein, on the other hand, only intended to make a coin-sized hole. Even so, Libet could stimulate the cortex immediately below this small entry point, and other cells situated below the surface could be stimulated by pushing the electrode down into the brain.

The hole that Feinstein made in the skulls of his patients was immediately above the sensory and motor cortex – that is the part of the cortex subserving touch and movement. By stimulating the sensory and motor cortex Libet was able to reproduce some of Penfield's findings. Patients reported a range of tactile sensations, some of which were vivid and detailed – for example a drop of water trickling down the back of the hand; however, Libet also noticed something very interesting. Most patients took approximately half a second to report sensations. Moreover, if Libet began stimulating the cortex but stopped before a half second had elapsed the patient would report nothing. These observations suggested that it took half a second for activity in the brain to reach awareness. Put another way, every experience is preceded by half a second of preconscious processing.

Although we experience the internal (mental) world and the external (physical) world simultaneously, this must be by virtue of some neurological trick. If the brain is a kind of biological machine, then it will take time to perform any of its functions – including the construction of consciousness.

In the 1840s Helmholtz had demonstrated that nerve impulses travel very quickly, but measurably (for example leg nerves carry signals at about 200 miles per hour). In fact, human nerve impulses travel at different speeds, mediated by numerous biological factors. Interestingly, although human beings have a tendency to self-aggrandise by citing 'the speed of thought' as a kind of competitor to 'the speed of light', nerve impulses in the brain rarely exceed relatively modest velocities. Consequently it must take time to build a mental representation of the world, and even longer to respond to that representation. Libet had simply confirmed what Helmholtz and common sense already suggested. Nevertheless, he had also advanced knowledge by establishing a specific temporal parameter, describing the half-second lag separating cortical stimulation and the entry of a tactile sensation into awareness.

In order to achieve a better understanding of the half-second of preconscious processing preceding awareness, Libet began to experiment with *evoked response potential* (or ERP) recording. The brain functions electrochemically, and for many years, scientists had recorded 'brain waves' by fixing electrodes to the scalp. This form of measurement is known as electroencephalography (or EEG), and produces a continuous record of wavy or jagged lines. ERP recording is a special form of EEG and was relatively new when Libet decided to use it in his programme of research.

ERP recording involves exposing an individual to the same stimulus hundreds of times. If all of the subsequent EEG lines are averaged, the result is a single brain wave. This curve represents a relatively pure response to the stimulus. Essentially, ERP recording is a way of getting rid of EEG signals that have nothing to do with the brain's specific response to the stimulus.

Libet recorded directly from the brain, placing his electrodes on the cortex in order to get a particularly good signal. The stimulus he used to provoke a response was a mild shock administered to the back of the hand. The subsequent brain wave showed a very early response, that is, within 10–20 milliseconds. However, the curve continued to rise, showing the brain in the process of building awareness. Libet had actually plotted preconscious processing on a graph – a shallow arc, climbing towards consciousness.

Libet's research had now raised some disconcerting philosophical issues. If consciousness arrives late (some half a second after related brain activity begins), how can human beings exercise free will? The decision to stand up, for example, must be preceded by a half second of brain activity that eventually results in that decision entering awareness. Libet's results suggested that all behaviour must – to a greater or lesser extent – be initiated unconsciously. Once again, the shadowy presence of a hidden intelligence seemed to be impressing itself on the debate, mischievously shaking the precocious pens of Libet's EEG machine.

By the 1970s ERP research conducted by the German neurophysiologist Hans Kornhuber and his assistant Lüder Deecke had demonstrated that voluntary actions were preceded by considerable electrical activity in the motor cortex, sometimes beginning up to a second before the action takes place. They called the subsequent ERP curve a *readiness potential* (as it showed electrical changes reflecting the brain's readiness to perform an action).

The discovery of the readiness potential revived Libet's interest in the relationship between volition and preconscious processing. Subsequently, he set about devising a task that could determine the exact moment at which experimental subjects decided to perform a simple action. Subjects were placed in front of a screen and asked to observe a circling spot of light, the trajectory of which followed a 'clock face'. After at least one revolution, subjects were instructed to move a finger (or hand), but in their own time. The decision was entirely theirs. When subjects felt the urge to move arise, they were asked to follow it through with an action, but to note the position of the spot of light (i.e. when they first experienced the urge to move). This allowed the onset of the urge to move to be established with millisecond accuracy.

The results of these experiments were consistent with Libet's earlier work. Subjects reported that they first became aware of the urge to move approximately 200 milliseconds before the action was executed; however, inspection of their ERP recordings showed that the motor cortex was busy planning the finger and wrist movements earlier. Indeed, the ERP recordings showed the now familiar half-second delay between the onset of brain activity and per-formance of the related action. The unconscious mind (or brain, at least) was definitely pulling the strings.

Although Libet's collected studies were now looking like a formidable and persuasive body of research, there were still those who were disinclined to accept that preconscious processing was taking place. They took the view that, at best, Libet's work merely showed that no significant processing could take place in the brain in under half a second. Subsequently Libet sought to demonstrate that subtle processing effects could be detected in periods of less than half a second, and to accomplish this he returned again to his early method of direct stimulation. Conveniently, by the 1990s there were people walking around who had already had electrodes permanently embedded in their brains. It was no longer necessary to spend hours in an operating theatre to undertake the experiment that Libet had in mind.

All Libet's subjects suffered from chronic pain. Electrodes had been embedded in their brains which, when activated, blocked pain signals. These therapeutic bursts of electricity were delivered by the patients themselves on an as-required basis from a power source worn on their belts. When Libet activated the electrodes for more than half a second, all of his subjects reported experiencing 'something'; however, activations of less than half a second were not noticed at all. Then, in a specific time interval communi-cated by a light display, Libet asked his subjects to guess whether or not their electrodes had been activated – even though they were activated for less than half a second (and therefore apparently undetectable). Remarkably, Libet's subjects guessed accurately about 65 per cent of the time, and some subjects guessed accurately when given as little as 150 milliseconds of electrical stim-ulation. This work suggested that the brain was capable of processing infor-mation in very short periods of time; if this was the case, it was very likely that the half-second delay separating the onset of an ERP and the execution of a behaviour was filled largely by preconscious processing. The brain was not just waiting for consciousness to arrive but was engaged – among other things – in the task of making consciousness.

Libet's studies span over thirty years and, taken together, represent an extraordinary and deeply perplexing contribution to neuroscience. His results seem to undermine completely the view that human beings can exer-cise free will. It would seem that everything a human being decides to do (be it tapping a finger or contemplating the existence of God) is determined by brain activity that precedes awareness by about half a second. Moreover, because it takes at least half a second to become aware of stimuli, con-sciousness is constantly trailing 'real time'. We are all living in the past.

This latter proposition has been the source of considerable debate. We have no sense of a half-second delay as we transact with reality. It seems implausible that our affairs are conducted outside of real time. To address this criticism, Libet suggested that the brain is capable of glossing over inconsistencies between real time and neurological time. An illusion is created in which the delay is simply edited out. Although we are unaware of preconscious editing, the brain does this all the time. Think, for example, of the blind spot – the oval-shaped area of the retina where the optic nerve joins the eyeball. It creates a six-degree island of blindness in the visual field, which is large enough to accommodate six full moons. This handicap is never problematic because the brain simply fills in the gap. Contemporary neuroscientists have suggested that the blind spot represents a fundamental example of an interpolative process, of which there are probably many, designed to create an illusory though practical correspondence between phenomenal and real worlds.

Another robust example of the brain's penchant for interpolation is the *phi phenomenon*. Two alternating light bulbs will produce the illusion of a single spot of light bouncing back and forth. (This principle is exploited to produce rotating displays around theatre billboards.) If two different coloured lights are used, for example red and green, and the alternation occurs within 50 milliseconds, the bouncing light appears to change colour halfway across. In other words, the brain perceives a colour change (in this case from red to green), before the green bulb has been lit.

Libet's suggestion that the brain is constantly glossing over discrepancies between real time and neurological time becomes considerably more plausible when placed in a broader context. It would seem that the brain is equipped with numerous Orwellian mechanisms, continuously engaged in the process of distorting and editing phenomena, such that human beings can comfortably occupy their existential niches.

Nevertheless, critics have continued to argue that Libet's position is untenable insofar as his findings do not correspond with everyday experience. If you accidently touch a hotplate with your finger, it doesn't take you half a second to pull away. If a child jumps out in front of a car, it does not take half a second for the driver to hit the brake pedal. How, then, does Libet explain such phenomena? In actual fact, he doesn't have to. Such questions are fundamentally flawed, confusing behaviour with consciousness. Human beings frequently react in under half a second to environmental demands, but such reactions are not accompanied by consciousness. Rapid reactions usually occur automatically, with consciousness arriving late. It would be extremely unusual to experience awareness of a burning sensation while one's finger was still in contact with a hot object. The finger has always been withdrawn by the time one is first aware of being burned. If anything, a moment's reflection on the actual experience of performing rapid reactions seems to reinforce rather than weaken Libet's position. Preconscious processing routinely influences behaviour in under half a second. In such circumstances, the role of consciousness is not to initiate action, but to make sense of it. Consciousness is like a forensic scientist who must necessarily appear at the scene of the crime after the event in order to collect evidence and draw conclusions.

The late arrival of consciousness can also be observed directly on waking from dreamless sleep. It is sometimes possible to recall an odd state of wakefulness – lasting approximately half a second – that precedes the appearance of identity. We seem to exist only as a primitive, neutral awareness, without autobiography. True consciousness must necessarily involve a sense of self, which presumably cannot be constructed in under half a second. Consequently we are able to experience work in progress – where the 'work' in question is ourselves.

A further aspect of Libet's work which deserves underscoring is that it decisively demonstrates the independence of registration and conscious representation. It is possible to stimulate the brain, confirm that something has been registered by inspecting the EEG, and then abort the process before conscious representation has been achieved. This is an important finding, as it suggests that low levels of stimulus intensity – as might be employed in subliminal perception studies – are probably strong enough to register in the brain without a corresponding event in consciousness. One of the principal arguments against subliminal perception has been that stimuli presented at intensities too low for conscious representation are simply not registered by the brain. Libet's work strongly suggests that this criticism is invalid.

Although Libet and Freud undertook very different enquiries into the nature of mental life, and employed very different methods, they arrived at similar (if not the same) fundamental conclusions: the exercise of free will is probably an illusion and the real determinants of human behaviour are unconscious. Because free will is such a cherished concept, Libet's findings were welcomed with the same chilly reception that had previously been reserved for Freud.

Libet has attempted to preserve some dignity for the human race by suggesting that his work does not dispose of free will entirely. For example it is possible to inhibit a behavioural response, even if the early stages of preconscious processing that will eventually engender that response have already been initiated; however, the power to veto decisions is a very diluted form of freedom and reflects, perhaps, an obligation felt by Libet to offer a shred of consolation to disgruntled humanity. The fact remains, if Libet's work provides an accurate picture of how brain activity and consciousness are related, we are constantly being hoodwinked by the unconscious into believing in a non-existent liberty.

Libet's early work proved that a concept associated with nineteenth-century psychology – the unconscious – could be profitably investigated within the framework of contemporary neuroscience. Within a few years another research programme was underway which would perform a similar feat with respect to the equally dated concept of dipsychism (i.e. belief in a secondary, unconscious ego).

The human brain has two hemispheres. Although they look the same, and contain identical structures, they are, in fact, differentially endowed. This asymmetry first became apparent through observation of patients suffering from localised brain damage. Injuries to the left side of the brain tend to affect linguistic ability whereas injuries to the right side of the brain tend to affect spatial ability (for example understanding diagrams or pictures). This asymmetry is sometimes – but by no means always – reversed in left-handed individuals.

The rough division of functioning observed in clinical settings was soon confirmed in the laboratory – and later refined. It seems that the left hemisphere is specialised with respect to the comprehension and expression of language and the performance of tasks requiring detailed analysis and logic. The right hemisphere is specialised with respect to processing visual and spatial information, but it is also equipped to perceive holistically. The right hemisphere may also possess abilities that could be described as intuitive.

The two hemispheres of the brain are connected at several points, but the most significant connection consists of a thick band of nerve fibres called the *corpus callosum*. The corpus callosum seems to be the principal conduit through which information is exchanged between hemispheres. In effect, it allows the two hemispheres to converse. Unfortunately, in certain forms of epilepsy, the corpus callosum also serves as a convenient bridge, facilitating the passage of abnormal electrical activity from one side of the brain to the other.

Such patients can benefit from an operation known as a *commisurotomy*, which involves severing the corpus callosum. This procedure confines abnormal electrical activity to the diseased hemisphere (thus diminishing or abolishing seizures). Between 1962 and 1968 nine successful commisurotomies were performed by Los Angeles neurosurgeons Philip Vogel and Joseph Bogen.

At first sight, Vogel and Bogen's patients seemed completely unchanged –

apart from the obvious improvement in their health. Even though the two hemi-spheres of the brain had been disconnected, these patients showed no loss of intelligence and no other signs of brain damage; however, they *were* changed. They now exhibited some subtle and curious processing characteristics that could only be observed in special laboratory conditions. Moreover, these pro-cessing characteristics were so unusual that some fundamental assumptions concerning the unity of consciousness were substantially challenged.

In order to appreciate fully the experimental procedures employed to study 'split-brain' patients it will be necessary first to describe some basic features of cerebral physiology.

The 'wiring' of the visual system is quite complex. Each side of each eye transmits visual information through nerve pathways to both sides of the cerebral cortex. The net result is that if you stare straight ahead, everything to the right goes to the left hemisphere and everything to the left goes to the right hemisphere. All visual information is then shared by transmission through the corpus callosum. This of course does not happen in split-brain patients. Providing they keep their eyes fixed on a point straight ahead, all the information from the left and right sides of the visual field will remain iso-lated in respective hemispheres.

The wiring that subserves physical movements is less complicated. The body is simply cross-wired in a straightforward manner. Thus, the left hemi-sphere controls the right side of the body and the right hemisphere controls the left side of the body.

With these facts in mind, we can now consider the remarkable set of experiments conducted by psychobiologist Roger Sperry and his colleagues at the California Institute of Technology. They were conducted on Vogel and Bogen's split-brain patients, all of whom were right-handed and showed con-ventional patterns of hemispheric asymmetry (i.e. their speech was located in the left hemisphere and their visuo-spatial abilities in the right hemisphere).

Patients were instructed to look straight ahead while visual information (i.e. words, drawings, and pictures) appeared very briefly on the right or left side of a display screen. Thus, stimuli appearing on the right were conveyed to the left hemisphere and images appearing on the left were conveyed to the right hemisphere. Each image only appeared for a twentieth of a second, which is long enough to ensure conscious registration but not long enough for the eyes to shift. If the eyes were allowed to move then this would result in stimuli being registered in both hemispheres. Sperry's interest was in how patients would respond to different stimuli presented exclusively to one or other side of the brain. Patients could respond verbally, or with their hands (although both hands could be hidden under a screen).

If a word was flashed to the left hemisphere patients could say what it was and write it with their right hand; however, if a word was flashed to the right hemisphere the patient could neither say what it was nor write it. This was a dramatic confirmation of how processing abilities are asymmetrically distributed in the brain – with language being located exclusively in the left hemisphere. Nevertheless, it was subsequently established that although the right hemisphere is mute, it can still recognise objects. Moreover, if provided with a method of communication that does not involve speech, it will make a correct identification without difficulty.

If a picture of an object, say a fork, was flashed to the right hemisphere and the left hand allowed to feel for a fork among many different objects, a correct identification was accomplished more or less immediately. Even so, having correctly identified the fork, patients could still not say what it was. If, however, the fork was then transferred from the left to the right hand (permitting the left hemisphere to make a contribution) patients were able to say 'fork'.

As predicted from clinical observations, the right hemisphere proved superior to the left hemisphere when tasks required visual and spatial processing. For example, arranging blocks and drawing cubes was achieved more competently with the left hand than the right, even though the patients were right-handed.

Sperry's results demonstrated that the corpus callosum was the means by which the two hemispheres of the brain exchanged knowledge – straightforward facts about the world. But what about emotional information? Did emotional information use the same channel?

Although the corpus callosum is severed in split-brain patients, the two hemispheres are not entirely disconnected. Beneath the corpus callosum another pathway – the *anterior commissure* – is preserved. These fibres evolved in the brain much earlier than the corpus callosum, and connect the structures located deeper in each hemisphere described collectively as the *limbic system*. The limbic system is associated with the processing of emotionally significant information, but not directly involved in representing such information in awareness. Expressed more simply, the limbic system operates unconsciously.

The response of split-brain patients to emotional stimuli was explored largely by neuroscientists Michael Gazzaniga and Joseph Le Doux. The results of their research programme suggest that emotional information flashed to the right hemisphere can provoke an emotional response that the left hemisphere experiences, but cannot explain. For example, when the image of a nude woman was flashed to the left hemisphere of a female patient, she laughed and correctly identified the image; however, when the same image was flashed to the right hemisphere the patient claimed to have seen nothing, but proceeded to chuckle anyway. When asked what she was laughing at, she said that she didn't know, and implicated the experimental apparatus, saying: 'Oh – that funny machine'. Such responses suggest a curious state of knowing, yet not knowing.

With the aid of specially designed contact lenses it became possible to expose stimuli for longer periods of time without risking information being registered in both hemispheres. Subsequently more powerful and complex emotional material could be used. Films of a disturbing scene shown to the right hemisphere could not be described, yet they reliably evoked feelings of fear and distress. Similarly, films of emotionally pleasurable scenes aroused congruent emotions, but were experienced as equally inexplicable to the patient.

These results indicate that pathways exist in the brain for the transmission of 'emotion', across the anterior commissure, from right to left hemispheres; however, in the absence of the more sophisticated transmission route of the corpus callosum, the left hemisphere is not equipped with sufficient knowledge to understand why the emotion has arisen. An everyday phenomenon that might reflect this experience is the appreciation of art. Many people respond viscerally to abstract works such as those by Mark Rothko and Jackson Pollock. The visual pattern is processed in the right hemisphere, but because there is no obvious meaning to communicate to the left hemisphere only an emotional response is produced. Apart from those with special knowledge of art, most people are familiar with the experience of being able to express unequivocal pleasure when viewing a painting while at exactly the same time being unable to say why. If pressed, one might 'make up' a reason – but ultimately such a reason can only be a post-hoc rationalisation. Interestingly, split-brain patients do much the same thing.

Gazzaniga and Le Doux noticed that their split-brain subjects were inclined to confabulate after emotional responses had been evoked by right-hemisphere presentations. Although they had no idea why they were feeling scared or happy, they would happily proffer an explanation. Such explanations were, of course, inaccurate; however, the ease with which they were generated suggested a further function of the left hemisphere – namely, to help shore up an illusory sense of authority. This function could represent yet another Orwellian process that helps human beings to accommodate anomalous experiences that must necessarily arise when a part (or parts) of the mind operate independently.

With respect to the original programme of split-brain research, Sperry concluded that surgery had left split-brain patients with 'two separate spheres of consciousness'. But to some degree, all human beings must possess two spheres of consciousness. The surgeon's knife merely emphasised existing partitions. Underlying all experience is a divisible neurophysiology. This suggests that the faculties of the mind are predisposed to separate more readily along certain biological fault lines: reason from intuition, unconscious emotion from conscious emotion, and so on. Moreover, the two hemispheres of the brain are so differently endowed, they might almost be construed as possessing distinct personalities. In the split-brain literature we witness the curious spectacle of contemporary neuroscience revisiting the nineteenth-century world of alter egos – Dostoevsky's *The Double*, Stevenson's *Jekyll and Hyde*.

But to what extent is such speculation valid?

In an article called 'The split brain in man' (1967), Gazzaniga wrote:

> All the evidence indicates that separation of the hemispheres creates two independent spheres of consciousness within a single cranium, that is to say, within a single organism. This conclusion is disturbing to some people who view consciousness as an indivisible property of the human brain. It seems premature to others, who insist that the capacities revealed thus far for the right hemisphere are at the level of an automaton.

Clearly, a definitive demonstration of two independent spheres of consciousness in the same cranium is made somewhat difficult by the absence of language in the right hemisphere. If the right hemisphere possessed the ability to express an opinion (however crudely) that differed from the left hemisphere, the argument for an underlying physiological basis for dipsychism would be considerably more compelling. The experimental requirement was a split-brain subject whose right hemisphere possessed significant language skills. Luckily, just such a subject appeared.

The patient, known as PS, was a sixteen-year-old boy who had been given a commisureotomy – severing the corpus callosum and separating the brain hemispheres – in 1975. Again, this was performed to treat his epilepsy. In order to ask PS's right hemisphere questions an elaborate experimental design was required. This was because spoken questions, unlike visual images, cannot be delivered to one or other hemisphere. The pathways that subserve the processing of auditory stimuli relay information to both hemispheres. Nevertheless, through an ingenious combination of spoken and visual presentations it was possible to pose a question to one or other of PS's hemispheres. Although PS's right hemisphere could not speak, it could express itself by arranging letters.

The results of the subsequent experiments were astonishing. It transpired that PS's right hemisphere could disagree with his left on a range of topics, from the mundane – such as colour preference – to more complex personal matters. The most dramatic demonstration of hemispheric independence occurred when PS was asked what occupation he intended to pursue after graduation. His left hemisphere wanted to be a draughtsman, but his right hemisphere wanted to be a racing driver.

A further interesting feature of PS's psychology was that his mood varied according to how well his two hemispheres were 'getting on'. Such a finding is curiously consistent with Freudian and Jungian ideas that equate psychopathology with conflict between mental agencies.

Recent scanning studies have also revived another psychoanalytic idea; namely, that early traumatic experiences can divide the psyche. Work undertaken at McLean's Hospital (a psychiatric centre affiliated to Harvard Medical School) has demonstrated that children who have been neglected or abused show a reduced corpus callosum size. On average, it is up to 40 per cent smaller. A consequence of this might be that the locus of brain activity mediating personality becomes relatively isolated in each hemisphere. Eventually, this might result in the development of two quite different facets of personality, either of which might become dominant in a given circumstance. For example, the McLean team have suggested that recollections of abuse will produce an emotional state signalling a predominance of activity in the right hemisphere.

If the principal pathway of communication between left and right hemispheres of the brain is compromised in abused children, then such individuals might enter adulthood resembling split-brain patients. An ordinarily articulate individual might find it difficult to find words when recollections of abuse have shifted the locus of brain activity into the right hemisphere. The vocabulary and linguistic skills of the left hemisphere might become less accessible – the result of which would be the emergence of what might appear to others to be a secondary personality.

Further evidence for a form of dipsychism rooted in hemispheric asymmetry comes from a bizarre neurological phenomenon known as *alien hand*, in which the non-dominant hand – usually the left – seems to be under the control of a mischievous alter ego. Alien hand occurs in individuals who have suffered a brain injury affecting the motor cortex or, predictably, the corpus callosum. Needless to say, the phenomenon has also been observed in split-brain patients.

The degree to which an alien hand can operate independently can be quite dramatic. For example, items can be selected while shopping that the buyer has no interest in. It is also typical for alien hands to 'behave' in a contrary fashion, reversing the actions performed by the dominant hand. For example, unbuttoning a shirt that has just been done up or, even worse, pulling down trousers that have just been raised.

It is customary for neurologists to underplay the psychological signifi- cance of alien hand phenomena. The accepted view is that the behaviour is akin to an unwelcome reflex. Nevertheless, it is tempting to endorse an obvious and less pedestrian explanation; namely, that damage to the corpus callosum allows an ordinarily incapacitated and unconscious alter ego to make its presence felt. Like an actor in a low-budget horror movie the unfor- tunate victim is – at least in part – taken over.

The opposite of alien hand is a neurological condition known as *neglect*. This is when brain injury results in a curious attentional problem in which everything to the left or right is simply ignored. It is as though one or other of the two hemispheres loses the ability to concentrate and retires from active service. Patients suffering from neglect might only eat half the food on their plates, only comb half their hair, or only manage to dress half their bodies. Moreover, they may only be able to turn left or right and, when initi- ating an action, may forget to engage half of their limbs.

Advances in brain-scanning technology have allowed neuroscientists to actually peer into the working brain. Typically, brain scans produce images in which areas of activity are shown by bright patches of colour. These bright patches represent physical changes such as increased blood flow or con- sumption of oxygen or glucose.

Scans of patients suffering from neglect show that the visual cortex is func- tioning normally. Although they are not conscious of stimuli presented in the blind half of their visual field, the information is certainly being processed.

These results would have been of considerable interest to cognitive psy- chologists conducting experiments on attention on the 1950s. Neglect patients exhibit an extreme and special form of 'not paying attention'. Modern brain scans convincingly demonstrate that unattended information is being analysed, if only for its basic physical features.

Brain-scanning technology has provided evidence for an even more dra- matic demonstration of information processing in the absence of awareness. Recently, a patient in a coma was shown photographs of familiar faces (by retinal projection). The scan revealed activity in cortical regions that 'light up' when faces are shown to the conscious brain. Clearly, even in the complete absence of awareness, the brain is capable of undertaking a feature analysis. And given that some patients are sometimes roused from comas by the sound of a familiar voice, who can really say to what degree the 'inert' brain is not capable of a more sophisticated, semantic analysis?

Perhaps the most celebrated example of unconscious perception is the phenomenon now known as *blindsight*. As the starkly oxymoronic name suggests, blindsight refers to the accomplishment of tasks associated with being sighted by people who, given their injuries, should be blind. The phenomenon has a relatively long history. For many years people working in close proximity with the blind had observed instances of oddly competent behaviour (for example blind individuals who seemed to have an uncanny knack of being able to side-step obstructions which would otherwise have caused them to trip over) and in the First World War blinded soldiers were reputed to have avoided ordnance by ducking – even though they seemed to have no idea what they were doing.

Blindsight only occurs in individuals whose blindness is due to damage affecting the visual cortex. The eyes must be unharmed. Thus, in cases of cortical blindness, visual information arrives for processing at the visual cortex but a conscious representation cannot be constructed. Cortical blindness can be total (where the entire visual cortex is damaged) or partial (which arises when a limited area of the visual cortex is damaged). In the latter case, one or several islands of blindness mottle the visual field.

The first person to undertake a systematic investigation of blindsight was Oxford psychophysiologist Larry Weiskrantz. In the early 1970s he began studying a patient who had sustained damage in an area of the visual cortex known as the right calcarine fissure. The result was that the patient was unable to see anything in the left half of his visual field.

Weiskrantz seated him in front of a screen on which different stimuli were presented – although all on the left. The patient was instructed to reach out and touch the screen when something appeared. This was, of course, a seemingly bizarre instruction, given that the patient couldn't see anything. Nevertheless, he co-operated – just guessing.

It transpired that the patient's guesses were remarkably accurate (which was as surprising to him as it was to Weiskrantz). Refining the procedure, Weiskrantz established that, although cortically blind, his unusual subject could indicate the position of a light flash, discriminate correctly between horizontal, vertical, and diagonal lines, and between the letters X and O. All presented in 'blind' regions of the visual field. The patient insisted that he had no awareness of the stimuli; however, he did admit to sometimes experiencing a 'feeling' that 'something was there'.

Since the early days of blindsight research more subjects have been examined and scanning technology has permitted a closer look at underlying mechanisms. Indeed, brain scans have shown that, although the principal area of the visual cortex does not 'light up' during blindsight, another small area – known as V5 – does. This is an area concerned with the detection of movement. It would seem, then, that images presented in blind areas do receive a very basic analysis; however, this analysis is not sufficient to produce conscious representations. Even so, a registration of some kind does occur, facilitating the production of relatively accurate 'guesswork'.

The concept of the cognitive unconscious that emerged during the 1980s – that is the concept described by the likes of John Kihlstrom and described in Chapter 6 – hinges on the notion of progressive automation. Thus, consciousness might be needed to co-ordinate and assemble a new behaviour (be it mental or physical), but with practice, diminishing resources are required. Once again, brain-scanning technology has exposed this process taking place.

For example, on certain language tasks (such as matching verbs with nouns), scanning investigations show a burst of activity affecting many disparate areas of the brain; however, within a few repetitions, areas of illumination shrink dramatically. Moreover, the areas of the brain associated with higher-order functioning – i.e. those that co-ordinate and supervise more simple processing areas – disappear altogether. One can actually observe the transfer of conscious to preconscious processing; the brain's habitual inclination to abbreviate everything.

These results are entirely consistent with the theoretical justification of the division between conscious and unconscious minds. If the brain can swiftly reduce tasks to 'habits', then more processing resources can be allocated for the analysis of ongoing environmental contingencies. In other words, the brain is in a better position to make plans, deal with new problems, and thus increase its own chances of survival.

Scanning investigations showing the brain converting conscious mental procedures into unconscious mental procedures demonstrate an appealing symmetry. Libet suggested that it takes the brain half a second to raise an event into consciousness. Brain-scanning studies show traffic flowing in the opposite direction. Although it may take somewhat longer, the brain is equally adept at lowering the contents of awareness into the unconscious. Automation is simply the conscious mind going unconscious.

In Freud's major 1915 essay 'The unconscious' he argued that physical theories of the unconscious represented something of a cul-de-sac. Neuroscience was insufficiently advanced to examine or explain the unconscious in terms of the physical characteristics of the brain. Since 1915 advances in neurosurgery, electroencephalography, and brain scanning have removed most of the limitations that turned Freud from a neurologist into a psychologist.

Although brain science and psychology are often and quite perversely presented as oppositional disciplines, they are, in fact, simply different ways of describing mental phenomena. Interestingly, contemporary methods of investigating the brain have not invalidated Freud's psychological unconscious; they have, if anything, added weight and depth to its foundations. Libet's EEG studies have gilded Freud's claim that free will is something of an illusion, and the subcortical structures that process information unconsciously are also those that generate our most id-like emotions. Scanners have penetrated the cerebrum and illuminated the dreaming mind. We can actually see the visual cortex constructing hallucinatory visions and the auditory cortex hallucinatory voices. Moreover, as dreams become nightmares, we can observe the deep structures that mediate fear flaring up like warning beacons. And all the time, while the dream is being orchestrated by subcortical structures and the sensory apparatus, the frontal regions of the brain – essential for testing the tensile strength of reality – glow with a much duller light. Just as the nineteenth-century opium addicts suggested, the conscious mind is anaesthetised, and must bare mute witness to the products of preconscious processing.

The most basic notion on which the unconscious depends is that the human mind can be partitioned. Clearly, split-brain patients show that an underlying physiology exists for several divisions, both horizontal and vertical. Not only can the intellectual upper storeys of the mind be separated from the emotional lower storeys, the vertical central sulcus which separates the two hemispheres shows a natural rift between two proto-identities. Thus, contemporary neuroscience invites comparison with ideas that pre-date Freud, and at first sight seem to be even more fanciful. The alter ego has wandered out of the late-nineteenth-century case book – the late-nineteenth-century novel – and found a new home in the modern laboratory.

Finally, scanning studies have even supported the cognitive unconscious, with its emphasis on automation. The shrinking areas of illumination observed on repetition of the simplest tasks are an elegant demonstration of the brain's propensity for heuristic tendencies.

One of the most potent criticisms of the unconscious has always been that it is not 'provable', not testable with instruments of science. In other words, not 'real'. The unconscious can only ever be a kind of metaphor – little more than a literary device.

In fact, the unconscious has proved to be as much a part of the physical world as granite or the sun.

Amnesia is often associated with head injury and occurs in two forms: *retrograde amnesia*, in which the patient's memory up until the head injury is erased, and *anterograde amnesia,* in which the brain loses the capacity to form new memories. With respect to the latter, intellectual functioning can remain relatively unaffected. Thus, the individual can remember everything he or she knew up to the time of the accident and will be just as intelligent.

Until recently it was assumed that patients with severe anterograde amnesia had lost the ability to learn completely; however, it has become increasingly apparent that learning does take place. Amnesic patients can learn new skills, new vocabulary, and new facts, but they cannot remember acquiring this information. Thus, their behaviour is influenced by knowledge which they claim they do not have. It is as though the unconscious is subtly and carefully guiding their choices and judgements while the patients themselves experience nothing more than guessing or following hunches. As such, amnesic patients provide a compelling example of the power of unconscious memories.

Although the influence of unconscious memories in the context of amnesia has only recently been the subject of careful scientific study, neurologists have always been aware that such phenomena occur. In 1911 the Swiss psychologist Edouard Claparède conducted a simple experiment on one of his amnesic patients which suggested she was making use of knowledge that she didn't know she had. Claparède concealed a pin in his hand while shaking hands with her. The following day, the same patient was reluctant to shake hands but could not explain why, opting instead for some perplexed confabulation.

Perhaps the most celebrated case of unconscious learning in modern neurology is the patient known simply as David – a man exhaustively studied by the neurologist Antonio Damasio and his colleagues. David is ostensibly unable to learn any new information. He has extensive damage to both temporal lobes of the brain, and the structures therein that serve memory.

Damasio designed an intriguing study to test David's implicit learning, which became known as 'the good-guy/bad-guy experiment'. Essentially, David was engaged in three types of social situation. The first was pleasant and rewarding, the second neutral, and the third unpleasant. In the third or 'bad-guy' condition, the individual engaging David assumed a brusque manner and administered a battery of 'psychological tests' designed to be dull and boring. These different types of interaction took place over five consecutive days.

David was then shown some photographs of faces, including those who had played the role of 'good', 'neutral', and 'bad' guy in the experiment. David was asked: 'Whom would you go to if you needed help?' And for further clarification: 'Who do you think is your friend from this group?' David chose the 'good guy' 80 per cent of the time. He chose the 'neutral guy' at chance levels, but almost never chose the 'bad guy'. Yet he claimed that he didn't recognise any of the faces he was shown and that he knew nothing about them. Clearly, learning had taken place; however, that same learning was not accessible to consciousness. Damasio explains David's behaviour by implicating unconscious emotions (i.e. subliminal feelings that are able to influence choices):

> He did not know why he chose one or rejected the other; he just did. The nonconscious preference he manifested, however, is probably related to the emotions that were induced in him during the experiment, as well as to the nonconscious reinduction of some part of those emotions at the time he was being tested.

Although David's unconscious learning is an intriguing phenomenon, it appears intriguing only because of its extremity. In actual fact, most people's day-to-day social preferences are probably influenced by unconscious learning too. It is only because the effect is diluted that it appears less remarkable.

Human beings are constantly forming positive or negative opinions of others, and often after minimal social contact. If challenged, opinions can be justified, but such justifications frequently take the form of post-hoc rationalisations. Some, of course, are laughably transparent. For example, 'He had funny eyes' or 'She had big ears'.

Unlike David, we can remember many of our past encounters. But, like David, there are many we have forgotten, and a forgotten memory is not necessarily a dead memory.

Just as the romantic poets and Freud suggested – the unconscious never forgets. When we take an instant dislike to someone, perhaps we are simply exhuming the prejudices of an unconscious that nurses its grudges in perpetuity.

8

Darwin in the dark

I sent Mr Darwin an essay on a subject on which he is now writing a great work. He showed it to Dr Hooker and Sir Charles Lyell, who thought so highly of it that they immediately read it before the Linnean Society. This assures me the acquaintance and assistance of these eminent men on my return home.

So wrote the young naturalist Alfred Russel Wallace to his mother on learning of how warmly his recently formulated theory of evolution had been received by Charles Darwin and Darwin's good friends Joseph Hooker and Charles Lyell. Without doubt, Mrs Wallace must have been very proud of her talented son. To have attracted the attention (and patronage) of such eminent men was a substantial achievement. This august trinity – Darwin, Hooker, and Lyell – seemed keen to facilitate young Wallace's ascent into the highest ranks of Victorian society. Or so it appeared. In fact, their behaviour was far from altruistic.

On 18 June 1858, Darwin's post included a manuscript written by Wallace outlining a theory of evolution. This in itself would not have rattled Darwin. Evolutionary theories had been in circulation for a very long time. The ancient Greeks Anaximander and Empedocles, their Chinese contemporary Tson-Tse, Darwin's grandfather Erasmus, Jean Baptiste Pierre Antoine de Monet (better known as Lamarck), and Darwin's contemporary Robert Chambers had all proposed evolutionary theories; however, they were all – to a greater or lesser extent – wrong. Wallace's theory, on the other hand, was right and almost identical to Darwin's.

Darwin had established the principle of evolution by natural selection many years before the arrival of Wallace's manuscript, although he had not made his views public. Having discovered what is arguably the single most important idea in the history of science, Darwin decided not to publish but instead to devote himself to an exhaustive study of barnacles which ran to several volumes. Clearly, he was a man disinclined to rush things. Unfortunately, his leisurely approach to publication had created a very disagreeable situation. Darwin was suddenly faced with a terrible moral dilemma. What should he do with Wallace's manuscript? It was self-contained, complete, and fit for publication. *The Origin of Species*, Darwin's own major work on the subject of evolution, was, on the other hand, still unfinished. If Wallace's manuscript was published, *The Origin of Species* would become nothing more than a very large addendum – a comprehensive and scholarly confirmation of Wallace's theory.

What was he to do? Or, more importantly, what *should* he do?

Charles Darwin was renowned for his decency. He was honest, scrupulous, and insistent on fair play. He was a man whose sense of propriety had earned him the respect of his peers and the admiration of his acquaintances. Darwin decided that, given his position, he couldn't possibly make a disinterested judgement concerning the fate of Wallace's manuscript. It would be improper for him even to try. Subsequently he wrote to his good friend Lyell (the country's most eminent geologist) and asked him to decide on an appropriate and equitable course of action; however, he also pointed out that – although he had been sitting on his own theory for over a decade – he was now just about ready to publish a 'sketch' of his own views on evolution.

Lyell consulted Hooker (a botanist later to be made director of the Royal Botanic Gardens), and it was decided that the work of both men, Wallace and Darwin, should be presented together at the next meeting of the Linnean Society. All were satisfied that the matter had been brought to a just and satisfactory conclusion. But in reality Darwin's dilemma had been resolved very much in his favour.

At that time, Wallace was not only completely unknown but also absent. He was a young naturalist, without reputation, still engaged on an expedition to the Malayan archipelago. Darwin, however, was a celebrated natural historian and acquainted with some of the most influential men of his age. These disparities between Wallace and Darwin are not insignificant.

Darwin's name was more memorable. Subsequently, it would be Darwin's name, not Wallace's, that interested parties would more readily associate with the new theory of evolution. Moreover, it was Darwin, not Wallace, who would shortly be in a position to follow the Linnean Society presentation with a magnum opus on the subject of evolution. Wallace had nothing like *The Origin of Species* languishing at the bottom of his cabin chest.

Had Alfred Wallace submitted his manuscript to a scientific journal, or indeed to any natural historian in the world other than Charles Darwin, we would now routinely employ the term Wallacism rather than Darwinism. But Wallace *did* send his manuscript to Darwin, and in doing so he lost his claim on posterity. For over a hundred years his name very rarely appeared in anything but footnotes and appendices.

Darwin's despatch of Wallace was an act of virtuoso Machiavellianism. While appearing to be entirely fair – an attribute much valued by Victorian society – he succeeded in sabotaging Wallace's career. Darwin had made no attempt to suppress Wallace's work. On the contrary, it appeared that Darwin had played a key role in ensuring Wallace's ideas were introduced to a large and distinguished audience. Of course, that's what Wallace (and his mother) believed; and in a sense, it was what Darwin believed too.

Although Darwin's treatment of Wallace was shabby, there is a considerable body of evidence to suggest that Darwin really was a very decent man. His correspondence written during the Wallace affair seems to show that he was determined to acquit himself honourably. So why did he sell Wallace short? One possibility is that – as far as he was concerned – he hadn't. In his view, sending Wallace's manuscript to Lyell had been an authentic, creditable act; however, given that Lyell would obviously make a decision favouring Darwin, it seems unbelievable that Darwin could have thought such a thing. Most people observing similar behaviour in everyday life would say of someone like Darwin: 'Well, he must be deceiving himself'.

Unfortunately, decency (even genuine decency) is not a straightforward matter. Decency can function as an extremely effective moral smokescreen; and the more genuine that decency appears, the easier it is to conceal a self-interested agenda. Being able to fool oneself has considerable advantages. If you can fool yourself, then you'll certainly be in a better position to fool others.

Darwin's real agenda was to dispose of Wallace. By sending Wallace's manuscript to Lyell he ensured Wallace's demise. But at the same time, he appeared to be selflessly promoting the interests of a would-be competitor. The fact that he accomplished this sleight-of-intention so brilliantly suggests that he must have been very convincing. And to convince others, he must have convinced himself. In other words, it served him well to be *unconscious* of his real agenda. It served him well to be ignorant of his own game plan.

In *The Moral Animal* (1994), the science writer Robert Wright has argued that during the Wallace affair Darwin dramatised two interesting implications of his own theory. Evolution endows human beings firstly with a powerful drive for self-advancement and secondly with an unconscious to help them practise the deceptions necessary to achieve self-advancement. The first point was well recognised by Darwin. Needless to say, the second point wasn't.

It would take over a hundred years for the links between Darwinism, deceit, and the unconscious to be properly understood; however, once these links were established the scientific legitimacy of the unconscious would be beyond question. Although the concept of the unconscious was of little interest to Darwin (or, of course, the unfortunate Wallace), evolutionary theory would provide a very convincing rationale for its appearance in human psychology.

Until relatively recently the relationship between psychology and evolution had attracted little interest. This is surprising, because Darwin himself was fascinated by psychology. For example, in 1872 he published *The Expression of the Emotions in Man and Animals*, a work which – as the title suggests – examined the then much-neglected topic of emotional expression. Darwin's thesis was that emotions could be understood in terms of their function or purpose. For example the posture, gait, and facial expression of an animal might communicate its intention to either submit or attack. This method of analysing behaviour is often described as functionalist. Within the context of an evolutionary framework, the functionalist approach suggests that, ultimately, any psychological phenomenon can be explained by identifying its survival value.

Psychology did not resist functionalism. Indeed, within a relatively short period of time psychologists were happy to acknowledge that all mental phenomena could be understood in terms of their purpose – or at least an original purpose which might have become somewhat obscured by more recent evolutionary developments. Nevertheless, functionalism did not change psychology. In textbooks functional considerations rarely developed beyond a superficial but respectful tug of the forelock in Darwin's direction. Typically a paragraph or two might recognise the functional significance of a mental phenomenon, but thereafter the discussion would more than likely proceed as if Darwin had never been born.

The theory of evolution by natural selection – to give the theory its full name – is based on a few simple tenets. Organisms produce more offspring than can survive and reproduce. Subsequently the organisms that survive tend to be better adapted. Parental characteristics appear in their progeny – thus, better adapted lines will survive and pass on their advantage. A few simple principles, but sufficient, nevertheless, to explain the provenance, appearance, and behaviour of all living things. Moreover, these few simple principles can also explain the emergence of consciousness and the division of the mind into conscious and unconscious parts.

It was work on the survival value of deception – conducted during and after the 1970s – that aroused general interest in the unconscious as an evolutionary concept; however, the unconscious had already been linked with evolutionary theory by those sympathetic to a functional perspective. It was assumed that the appearance of the human mind owed much to evolutionary pressures. Moreover, the moiety of conscious and unconscious functions could be understood as a kind of optimal solution to the problem of survival.

Clearly, the ability to reflect on need states, respond to novel stimuli, and make plans are all of considerable value with respect to ensuring survival. They are also functions for which consciousness is ideally suited; however, to execute these functions effectively, consciousness must also be strictly limited. The conscious representation of too much information – for example conflicting needs, unnecessary detail, or too many potential courses of action – would be counterproductive. The organism would be overwhelmed with information and paralysed by indecision. Therefore, all incoming information must be subject to at least a preliminary analysis before being selected for further processing and conscious representation. Moreover, such preliminary analyses must necessarily take place outside of awareness (if consciousness is to remain uncluttered). Finally, new skills, which can only be acquired by conscious rehearsal, must be automated as soon as possible; again, this feature of the processing system also serves to keep consciousness available for the performance of its primary functions.

In sum, evolutionary pressures explain why consciousness has arisen, why it is of limited capacity, and why it is complemented by processing activities that take place outside of awareness. The marriage of conscious and unconscious processing capabilities optimises the brain's capacity to negotiate complex and potentially hostile environments, ensuring survival and subsequent reproductive success.

The processing architecture of the brain has been shaped by evolutionary demands. It is a response to the world – an adaption; however, it is also a record. To peer into the brain is to look backwards in time. The vestiges of our remote ancestry are better preserved in the brain than in any natural history museum.

In the 1960s the neuroscientist Paul MacLean suggested that every human possesses not one brain but three: a reptilian brain, a mammalian brain, and, finally, what might be described as a human brain. The reptilian and mammalian brains recapitulate aspects of our remote evolutionary history. For example the reptilian brain – which is simple, primitive, and swells out at the top of the spine – reflects a stage in our history that pre-dates the development of emotion. The mammalian brain – which is more sophisticated and consists of a large system of linked structures – reflects the evolutionary stage of development during which emotions were established. The most highly developed region of the brain is the cortex – the outer layer that permits human beings to perceive complex stimuli, think through problems, and use language. The cortex is the most recent evolutionary development;

however, it is only a few millimetres thick. In physical terms, we are much more lizard and lion than Socrates or Einstein.

Much of what enters awareness is primarily the product of activity in cortical circuits of the brain. The deeper and more central areas of the brain – those reflecting our ancient history – operate beyond awareness. The unconscious foundations of human mental life were laid down in the brain while our very early ancestors were still dodging the tread of dinosaurs.

With respect to ensuring survival, one of the most important features of the mammalian, unconscious brain, is its potential to generate emotions – of which the most fundamental is probably fear. Contemporary neuroscience suggests that the conscious experience of fear is a relatively minor consequence of activity in a collection of structures that operate largely outside of awareness – the limbic system. It is the limbic system that generates basic emotions. When threatened, an organism equipped with a limbic system will respond very quickly, as a series of automatic 'programmes' are engaged. These produce the well-known 'fight/flight' response in which adrenalin is released, the heart beats faster, and muscles receive more nutrients owing to increased blood flow. Thus, the chemical environment in the body is altered to facilitate a speedy escape, or if this isn't possible, an energetic confrontation.

The neuroscientist Joseph Le Doux has identified the specific pathways and brain structures that subserve the fight/flight response. When a frightening stimulus – let's say a snake – enters the processing system, relevant information is sent down two neural pathways. The first of these pathways transmits information to the visual cortex via a structure called the *thalamus*, which then constructs an image of the snake. This image is merely a descriptive picture – a neutral representation of the snake as an object. Recognition areas then identify this object as a snake by retrieving information stored in long-term memory. Together the image and relevant information produce conscious awareness of the snake as a dangerous animal, and a structure called the *amygdala* is activated. The amygdala then sounds the alarm, thus initiating the fight/flight response.

This pathway, which results in the conscious recognition of dangerous objects, is relatively slow. It is certainly too slow to guarantee survival. By the time it has confirmed that the long, coiled object is indeed a snake, the snake might have already struck. In order to overcome this problem, the brain also possesses a quick and dirty response system. This second pathway links the thalamus directly to the amygdala. A crude representation of the snake (described by Le Doux as 'almost archetypal') is transmitted to the amygdala in a matter of milliseconds. Subsequently the fight/flight response is engaged before consciousness is engaged.

Of course, the late arrival of consciousness will provide a more thorough analysis of the 'long, coiled object'. When this is completed, it might transpire that the amygdala has sounded a false alarm. The snake might turn out to be nothing more than a length of rope, in which case this information is relayed to the amygdala, dampening the fear response. Needless to say, because the fear response is primitive and automatic, the fear programme may still run for a while, keeping the individual in a tense and alert state.

The reason why evolution has equipped the brain with a swift (although not necessarily accurate) processing system is quite straightforward. In the ancestral environment, predators struck quickly. Therefore, it was well worth responding to any potential threat – even if the majority of responses proved to be unnecessary. The cost of failing to respond (or responding slowly after careful, conscious, consideration) was probably death. On the other hand, there were no costs attached to being jittery. In the ancestral environment, it didn't pay to be cool.

Because the quick and dirty pathway is so important, it operates automatically. Over millions of years natural selection has effectively taken control away from the individual. When in a life-threatening situation the individual cannot be trusted to respond appropriately. He or she might dither and so die. When we startle, we are exhibiting ancient wisdom – the distillation of our species' learning history.

The conscious and unconscious pathways involved in the generation of fear are clearly complementary; however, with respect to avoiding opportunistic predators, the unconscious pathway is easily the more valuable. This consideration has prompted some theorists investigating the neurophysiology of fear to argue that (at least with respect to avoiding predation) the role of consciousness may be extremely limited. Indeed, consciousness may have evolved merely to check – retrospectively – the operation of numerous unconscious processing systems that constantly manage the organism's relationship with the environment.

This might be the reason why simple phobias are not amenable to reason. Most arachnophobics are perfectly aware that most spiders cannot harm them. Even so, they will still experience extremely high levels of fear when confronted by the most minute spider. Because consciousness is only loosely tethered to the underlying apparatus that generates fear, the intellect cannot be brought to bear on the problem. The unconscious simply won't listen, preferring instead to abide by the time-honoured adage 'better safe than sorry'. Individuals suffering from phobias may have inherited a first-class alarm system, but it is one that has now become something of a handicap in the controlled, protected environment afforded by civilisation.

In the modern world, power points and light sockets are ever present and potentially lethal; however, there are hardly any plug- or light-socket phobics. On the other hand, there are millions of individuals who fear harmless spiders, rarely encountered snakes, and domesticated dogs. The things that we fear most reflect our evolutionary history. We are still inclined to avoid those things that were best avoided on the plains of Africa. It would seem that we are born with nervous systems that have been 'wired' by natural selection to fear revenants from the ancestral environment – or at least wired *to learn to fear* revenants from the ancestral environment. For example, it might be that we are predisposed to acquire our own fears by observing the expression of fear in significant others (i.e. parents and peers). Thus, ancient fears could be preserved by a form of cultural transmission in which emotional distress is more readily associated with certain creatures.

This evolutionary bias in the incidence of phobias is called *preparedness*, and was first described by the psychologist Martin Seligman in the 1970s. Subsequent research has suggested three classes of stimuli that human beings are *prepared* to fear. Firstly, virtually anything that might be linked with disease. This is a very broad category, and may include substances that are perceived as dirty as well as certain creatures associated with decay and decomposition (e.g. maggots and worms). Secondly, animals that are either venomous or predatory. Thirdly, unfriendly *conspecifics* (i.e. other human beings whose facial expressions communicate hostile intentions). This latter category appears because human beings are social animals who live in dominance hierarchies. As such, it is as important to respond to threats arising from within the social group as it is to those arising from outside.

Even within these three categories, more precise predictions can be made concerning the potential of a particular stimulus to provoke fear. For example, with respect to animals, reptiles should evoke the greatest fear, because reptiles were the greatest threat to our very early mammalian ancestors. Reptiles are the archetypal foe. Perhaps this is why science-fiction B movies are so frequently populated by latex saurians. When we look at a snake and shiver, we do so because of an unconscious 'memory' that is some 224 million years old.

The preparedness hypothesis is well supported by studies employing Pavlovian conditioning. A neutral stimulus will eventually trigger fear if it has previously been paired with a negative stimulus (for example, an electric shock). It is easier to condition fearless subjects to show signs of fear when presented with pictures of snakes or hostile faces than it is with pictures of flowers or happy faces. Evolution has biased the associative machinery of the brain in order to increase the organism's chances of survival.

The Swedish psychologist Arne Öhman has conducted a number of studies showing that prepared stimuli, even when presented outside of awareness, can initiate the fight/flight response. This is particularly evident in individuals for whom the presence of a phobia suggests an underlying neurophysiology sensitive to prepared stimuli. Elevated levels of arousal were detected by monitoring the skin conductance response (SCR), which reflects very subtle changes in sweat gland activity (see Chapter 6). Arachnophobics, ophidiophobics (those who fear snakes), and non-fearful subjects were shown pictures of spiders, snakes, flowers, and mushrooms. Although the pictures were shown for 30 milliseconds, they were rendered subliminal by the subsequent presentation of a masking stimulus – a second and more powerful stimulus that blocks awareness of the preceding exposure. None of the subjects could consciously identify any of the images appearing before the mask. Even so, ophidiophobic subjects showed elevated SCRs specific to snakes, arachnophobic subjects showed elevated SCRs specific to spiders, whereas non-phobic subjects showed the same level of response to all stimuli.

Subliminal presentations of prepared stimuli can also trigger the fear response in non-phobic subjects – albeit after conditioning. Öhman and his colleagues paired pictures of prepared and control stimuli with electric shocks before presenting them to subjects again in a masked form to ensure subliminality. Subjects conditioned to snakes or spiders, but not those conditioned to flowers or mushrooms, showed elevated SCRs. The same effect was achieved with angry and happy faces. After conditioning, elevated SCRs produced by angry faces survived masked presentation while those produced by happy faces did not.

In the ancestral environment, a lone human was a dead human. Social animals can only survive in a social context. Even today, several million years on, things haven't changed a great deal. Those who are unwilling to accept, conform, or find a role within existing social structures will soon lose the benefits of group membership. Those who cannot properly assimilate within existing social structures are destined for a life of vagrancy and ill-health. The presence of welfare institutions will do little to mitigate the fate of those who don't 'fit in'. From a Darwinian perspective, their reproductive success is severely compromised.

Negotiating the social environment is largely about recognising and understanding others. This is why face perception is so important to humans. When we want to hide our own intentions we instinctively look away. When we want to discover someone else's intentions we look them in the eye. All hierarchical systems are by their very nature political. Therefore, the complicated business of establishing and maintaining one's position in the hierarchy necessitates an exquisite sensitivity to facial expressions.

Face recognition is the result of activity along two neurological pathways – one cortical and the other subcortical. The subcortical pathway travels through limbic structures (including the amygdala), and seems to invest images with emotional meaning before they are shunted back to the cortex. This complementary (rather than secondary) route – mediated by the unconscious brain – is strongly associated with emotional aspects of identification. Functioning independently, the limbic pathway could not perform the processing tasks that would result in face recognition; however, when it operates in tandem with the cortical pathway, perception is coloured by feelings of familiarity. The subcortical pathway operates more swiftly than the cortical pathway, which can result in an annoying feeling that one knows somebody without being able to say who they are. Full identification has to wait until the cortical pathway has completed its checks and is fully engaged.

This common experience shows that although the two recognition pathways are part of the same perceptual system, they can operate independently. Moreover, dissociations of a more extreme nature can be observed in individuals suffering from certain neurological conditions.

Prosopagnosia is an inability to recognise faces. Afflicted individuals may function well in all other respects, but be completely unable to distinguish one face from another. Prosopagnosia tends to be the result of damage affecting the temporal and occipital areas of the cortex (i.e. the lateral and posterior surfaces of the brain). Although the cortical processing areas are damaged, the subcortical processing areas may remain intact. When patients with prosopagnosia are shown photographs of faces – among which are included images of individuals who are familiar – they typically claim to be unable to recognise any of them. Nevertheless, when familiar faces are shown, prosopagnosic patients produce elevated SCRs. The subcortical pathway is able to classify the image as emotionally relevant, but the products of this analysis never reach consciousness.

Two other conditions demonstrate what happens when both pathways are contributing to the process of face perception, but the subcortical pathway is either over- or underactive.

When patients suffer from the *Fregoli delusion* they mistake strangers for familiar individuals – even though they are physically dissimilar. Typically, a patient suffering from this problem will wrongly identify a stranger as a friend or relative in disguise. He or she might also suggest reasons why the friend or relative has adopted a disguise, and these reasons often have a paranoid quality.

This situation is reversed when individuals suffer from the *Capgras delusion*. Afflicted individuals come to believe that familiar people have been replaced by imposters who have somehow assumed an identical appearance. The explanations that Capgras patients offer to justify this belief can be highly imaginative, involving anything from invading aliens to android replicas. When patients suffering from the Capgras delusion are shown photographs of familiar faces, they do not produce elevated SCRs. This underscores the importance of the emotional component of face perception. A fully conscious representation of a face is not sufficient to ensure accurate recognition. It must also be bathed in the half perceived, semi-conscious emotional emanations that rise from the deeper structures of the brain.

It is interesting that both Fregoli and Capgras patients exhibit post-hoc rationalisations (i.e. an individual is in disguise or has been replaced). Again, these 'explanations' may represent yet another example of consciousness trying to make sense of experiences that are largely determined by unconscious (or mostly unconscious) processes.

It is very possible that 'first impressions' are greatly influenced by subcortical rather than cortical pathways. The faces that we see at a cocktail party are evaluated with much the same neural equipment that was available to our apelike ancestors. When – for no apparent reason – we get bad feelings about people, or decide that we would prefer to avoid them, this is probably because of processing that has taken place beyond awareness. The emotional brain is warning us of potential sources of threat so that we can protect our position in the social hierarchy.

Social anxiety (which can sometimes reach phobic intensity) is probably caused in part by a general over-sensitivity to 'prepared' social stimuli. It may be that socially anxious people selectively attend to threatening faces, causing them to feel uneasy and insecure in social situations. Laboratory evidence supports this view, and shows that such attentional biases operate (at least initially) outside of awareness.

British psychologists Karin Mogg and Brendan Bradley showed pictures of threatening and non-threatening faces to anxious and non-anxious subjects. The faces were exposed for 14 milliseconds and rendered subliminal with a masking stimulus. Faces were replaced immediately by the appearance of a dot, to which all subjects were instructed to make a response as swiftly as possible. It was predicted that anxious subjects would react more swiftly to the dot when it replaced images of threatening faces (compared to non-threatening faces). This was because threatening faces should attract greater processing resources, thus facilitating awareness of the dot and accelerating subsequent reaction times. Mogg and Bradley's results were entirely consistent with these predictions. Anxious subjects responded faster to the dot when it was preceded by a subliminal threatening face. Moreover, supplementary data analysis showed that this effect was enhanced if threatening faces preceded the dot when it appeared in the left half of the visual field. This study not only confirmed the operation of an unconscious monitoring system for prepared social stimuli, but also suggested that the neural machinery subserving this monitoring system is located principally in the right hemisphere of the brain (the hemisphere generally associated with more instinctive feelings and appraisals).

Although avoiding threat is enormously important in evolutionary terms, the great value placed on survival is much more to do with optimising reproductive success than preserving personal safety. The ultimate reason why efficient harm-avoidance systems have evolved is to protect the genetic legacy. Given this fact, it is probably the case that a preconscious attentional bias for threatening faces is complemented by an equivalent bias for attractive faces. The oft-heard claim of ladykilling Casanovas and Lotharios that they are equipped with 'radar' for 'a pretty face' is probably more than mere braggadocio. Moreover, the ease with which attractive women can capture the attention of a passing male (even when occupying a peripheral position in the visual field) has rendered 'head-turner' a synonym for beautiful.

Again, because reproduction is so important in an evolutionary context, many of the processes that underlie sexual attraction operate automatically and beyond awareness. It is, for example, commonly believed that men find women with thin waists more attractive than women with wider waists; however, empirical research suggests this is not the case. It has now been established that men rate women more attractive – irrespective of weight – if they possess a low ratio of waist-to-hip measurement (i.e. possess a waist that is thin relative to the hips).

Most men are completely unaware that their attractiveness judgements are influenced by this ratio, believing instead that they are simply demonstrating a general predilection for slim women. In fact, slim women who do not possess a low waist–hip ratio are perceived as less attractive than larger women who do. It is very possible that evolutionary processes have shaped this unconscious preference in men for a simple and obvious reason – broad-hipped women are better equipped to give birth, so are superior vehicles for the safe transmission of genes into the next generation.

Consideration of the relationship between evolutionary pressures and human behaviour is by no means new. Darwin himself was always attempting to discern the evolutionary provenance of everything from kissing to apoplectic rage; however, the long courtship of evolutionary theory and psychology has only recently resulted in marriage. Prior to the 1990s the term 'evolutionary psychology' was not in common usage. Now it is routinely employed to describe a burgeoning discipline that has attracted considerable interest.

Psychology has always been a fragmented subject. Indeed, one of the major criticisms of psychology is that – unlike the natural sciences – it lacks broad, unifying theories. However, with the advent of evolutionary psychology it is now possible to understand many features of human experience and behaviour within a single impressive framework.

One of the least-expected consequences of evolutionary psychology was renewed interest in the unconscious. Even less expected was the inspiration for this renewed interest – the study of camouflage phenomena.

In the natural world, disguise is a popular survival strategy. Sometimes it can be extremely useful to look like something else. Thus, flowers can acquire the appearance of poisonous insects; butterfly pupae can acquire markings that resemble a snake's head; and so on. Such bluffs are very effective adaptations, and under specific environmental pressures can reach an extraordinary degree of sophistication. For example the Heike crab – which lives in Japan's inland sea – has 'learned' to disguise itself as a Samurai warrior. After a naval battle in 1185, in which many Heike Samurai were killed, a folk-legend developed concerning the fate of the survivors. It was believed that they had found a new home on the seabed, having magically assumed the form of crabs.

Today most Heike crabs possess shells that seem to be impressed with the face of a Samurai warrior. It seems very likely that this phenomenon has arisen because of the behaviour of successive generations of superstitious fishermen. Crabs whose carapaces were impressed with markings suggesting a human face were thrown back in the sea; whereas crabs sporting an ordinary carapace were simply eaten. Thus, fishermen have been conducting an informal selective breeding programme for almost a thousand years. The more a crab's carapace resembles a human face, the greater its chances of survival – and the greater its chances of passing on the genes bestowing this advantage to its offspring.

Deception is an effective survival strategy – but only to the extent that potential predators are genuinely fooled. Lying around pretending to be a flower is tantamount to suicide if your carnivorous enemy has worked out how to tell the difference. The nature of evolution is competitive and fluid. Thus, inevitably, predators develop improved methods of detecting deception, placing the onus on prey to respond by developing better methods of disguise. The situation is akin to an arms race, where even tiny advances in technology can alter the balance of power.

The kinds of deception observed in the animal kingdom are reflected (albeit more subtly) in human social behaviour, where the ability to dupe others is frequently practised in the service of self-interest. In humans, however, deception tends to take the form of psychological rather than physical disguise. But this strategy is fraught with problems. Contrary to popular opinion, human beings are – on the whole – quite genuine. They find assuming a psychological disguise extremely difficult, and even the simplest ruse can be ruined by unexpected and major complications. Faking a smile, for example, is far more difficult than would first appear. Indeed, in actuality, a fake smile can only ever be a pale imitation of its genuine counterpart.

A real smile (evoked by an authentic, positive emotion) requires the combined contraction of two muscle groups: the zygomatic major and the orbicularis oculi. The first affects the appearance of the mouth while the second affects the appearance of the eyes. Although it is possible to activate the zygomatic major by an act of will, the orbicularis oculi are not under voluntary control. They operate automatically, in association with activation in the limbic system. When we describe a smile as heartfelt – or coming from the heart – what we really mean is that is coming from the unconscious.

A false smile only engages the zygomatic major. It is, in effect, only half a smile. This is why false smiles never achieve the desired effect. Instead, they produce a strange, rather disturbed expression suspended somewhere between surprise and terror – the all too familiar blight of most group photographs. Needless to say, counterfeit smiles are unconvincing, and very easy to detect.

There are many other indices of insincerity. For example, lying is characteristically associated with fidgeting and a rise in vocal register. Moreover, frame-by-frame analysis of individuals known to be suppressing an emotion, reveal fleeting expressions that reflect the individual's real mental state. These so-called micro-emotions simply break through the façade. They cannot be held in check. The inadequacy of human imposture was underscored by Freud in one of his most famous observations:

> He that has eyes to see and ears to hear may convince himself that no mortal can keep a secret. If his lips are silent he chatters with his finger-tips; betrayal oozes out of him at every pore.

Clearly, the ability comprehensively to deceive others is a talent that can be readily exploited by a social animal; however, such exploitation is constantly associated with the risk of discovery – which, in turn, could prove very costly. To incur the anger of the group might result in expulsion – and as we have already established, in evolutionary terms, expulsion is a disastrous outcome for a social animal.

Were it possible to ensure a good performance, that is to say, were it possible to stem the flow of tell-tale signs that give the deceiver away, then deception would be a much more efficient method of optimising self-interest. But how might this be achieved? How is it possible to produce a genuine, heartfelt smile while plotting the demise of a potential rival? To do so, it would be necessary to conceal one's real intentions, not only from others but from oneself.

The idea that evolution might have equipped social animals with just such a mechanism for self-deception was originally proposed by sociobiologists Richard Alexander and Robert Trivers in the 1970s.

In his foreword to Richard Dawkins's book *The Selfish Gene* Trivers noted that if:

> deceit is fundamental to animal communication, then there must be strong selection to spot deception, rendering some facts and motives unconscious so as not to betray – by the subtle signs of self-knowledge – the deception being practised.

In other words, we deceive ourselves in order to deceive others better. Evolution favours a really good liar.

In a social hierarchy, the individual is constantly protecting his or her position while simultaneously monitoring the environment for opportunities that might lead to advancement. If the group's essential resources are limited, securing position and advantage might ultimately be a matter of life or death. In the ancestral environment – as in contemporary society – those at the bottom of the ladder suffer most during times of material hardship.

To optimise one's chances of survival in a social group, a certain amount of scheming becomes absolutely necessary. For example, it would appear that human beings have an innate tendency to make themselves look good, but at the expense of others. Moreover, they perform this manoeuvre outside of awareness in order to preserve the esteem of their peers and avoid discovery.

In 1978 psychologists Ruben Gur and Harold Sackheim published an article titled 'Self-deception: a concept in search of a phenomenon'. It contains details of a now famous experiment much cited by evolutionary psychologists, demonstrating how unconscious mechanisms mediate the relationship between self-esteem and self-deception.

When people hear a recording of their own voice they usually produce an elevated SCR. Because we are accustomed to hearing our own voices through a vibrating skull, most people find it quite difficult to establish ownership of a recorded presentation. Nevertheless, even when people are unable to identify their recorded voice, an elevated SCR shows that at some level, unconscious recognition has occurred.

If self-esteem is manipulated, for example by contriving repeated failure on a laboratory task, subjects are more likely to deny that a recorded voice is their own (even though they are still producing elevated SCRs); however, if self-esteem is raised (by contriving repeated success on a laboratory task), they are more likely to claim ownership of their own voice and, further, will claim ownership of voices belonging to other people. Once again, the unconscious seems to be cognisant of the real state of affairs, because the recorded voices of other people fail to produce elevated SCRs.

Reviewing this experiment, Trivers noted that human beings almost expand themselves when succeeding, but shrink themselves when failing. Success inflates the ego. We are happy to make ourselves more conspicuous when we are winning; when losing, we are inclined to lie low in the hope that our poor performance might more readily be overlooked. 'Yet', Trivers wrote: 'we are largely unconscious of this process'. The way we try to play up our successes and play down our failures escapes conscious representation.

What implications do these ideas have for consciousness? Again, the importance of consciousness appears to be much reduced. In a social-evolutionary context, the role of consciousness is simply to execute some slick PR work, while the real business of the day is conducted elsewhere, namely, outside of awareness.

Robert Wright has summarised the position very succinctly in *The Moral Animal* (1994):

> ... not only is the feeling that we are 'consciously' in control of our behaviour
> an illusion ... it is a purposeful illusion, designed by natural selection to lend
> conviction to our claims.

Once again, we hear a late reverberation that seems to owe its origin to Freud's mighty 'third blow': consciousness has been very much overrated and our behaviour is determined largely by unconscious processes.

It should be noted that this view is not inconsistent with a prior assertion; namely, that evolutionary pressures encouraged the emergence of consciousness. Consciousness allows the organism to reflect on need states, cope with novelty, and plan ahead. Nevertheless, in a social context, such functions may be secondary to image management. In a social context, consciousness is little more than a mask.

Evolutionary theory – particularly in the guise of evolutionary psychology – has had an almost incalculable impact on the study of conscious and unconscious processes; however, Darwin's ideas have had an even more profound effect on neuroscience. It was originally Herbart who suggested that ideas compete with each other for entry into awareness; however, this notion of Darwinian style competition has recently acquired considerable currency. For example, the neurophysiologist William H. Calvin and the biologist turned neuroscientist Francis Crick have both proposed theories of consciousness in which competition between nerve circuits is given central importance; however, the contemporary doyen of this approach is Gerald Edelman, whose general framework is tellingly described as *neural Darwinism*.

Edelman has suggested that natural selection operates within the brain. Thus, sensations have to vie with each other for space on the brain's mapping surfaces. Networks of cells are continuously competing with each other – with only a limited number of networks gaining sufficient territory to produce an event in consciousness. Thus, each moment of consciousness is a temporary resolution, punctuating a process of perpetual internecine struggle.

An attractive feature of Edelman's theory is that it explains the structure of the brain and the emergence of consciousness using the same set of principles. Each moment of consciousness is an adaption to environmental demands, in the same way that the brain itself is an adaption to environmental demands. With respect to the former, demands operate over a time-course of milliseconds; with respect to the latter, demands have operated over a time-course of millennia. Thus, Edelman's approach is truly unitary, suggesting evolutionary continuities between every fleeting moment of consciousness and the brain's ancient history.

Darwinian processes – whether they unravel in a single moment or over a period of a million years – are blind and unconscious. This was a sticking point for many of Darwin's critics. Complex structures such as the eye, the ear, or the brain appear to bear the hallmarks of design; however, they are in fact the result of natural selection and the random mutation of genes. Evolution is not guided by divine will.

In *The Blind Watchmaker* (1986), Darwin's contemporary champion Richard Dawkins considers the work of the Victorian theologian William Paley, who famously argued for the existence of God by underscoring the seeming implausibility of naturally occurring complexity. Paley pointed out that if, while walking on a heath, one discovered a stone, it would not be absurd to conclude that it had laid there for all of time. On the other hand, if one found a watch – a thing of complexity – one would automatically assume that it had been made by someone. Because the living world is filled with a variety of organisms showing much greater complexity than mere watches, Paley asserted that we are more or less obliged to accept that they, too, were made by someone.

The Darwinian response to Paley is articulated by Dawkins with ruthless eloquence.

> All appearances to the contrary, the only watchmaker in nature is the blind forces of physics, albeit deployed in a very special way. A true watchmaker has foresight: he designs his cogs and springs, and plans their interconnections, with a future purpose in his mind's eye. Natural selection, the blind, unconscious, automatic process which Darwin discovered, and which we now know is the explanation for the existence and apparently purposeful form of all life, has no purpose in mind. It has no mind and no mind's eye. It does not plan for the future. It has no vision, no foresight, no sight at all. If it can be said to play the role of watchmaker in nature, it is a blind watchmaker.

If Edelman is right, then the blind watchmaker is also at work in the brain. The same 'blind, unconscious, automatic process' that can produce marvels like a tiger or a whale can also produce the marvel of consciousness.

In later life, Alfred Russel Wallace's faith in evolutionary theory began to flag. Crucial to this change of heart was his consideration of human psychology. Wallace had observed that people seem to be endowed with gifts that have no obvious use with respect to survival. For example, he noted that 'savages' could be taught to perform calculations and play western musical instruments. These skills were of no use to them in the jungle. So why were they there?

Perhaps, Wallace concluded, our psychology is divine. Natural selection shapes the body, and the brain, but not the human mind. In the human mind, we see a pale reflection of the mind of God. Although Wallace always maintained that mankind had physically evolved from animal ancestry, he settled on the belief that our higher powers have a supernatural provenance. In the 1860s, he turned to spiritualism.

Had Wallace fully understood the power of evolutionary principles, he might have been able to deduce their impact on human psychology and the development of the human mind. More importantly, he might have been in a position to deduce their impact on the mind of Charles Darwin. Finally, he would have appreciated why it was that when he came to write his own book on the subject of evolution, he titled it *Darwinism* – not *Wallacism*.

9

The uses of darkness

In 1990, something unprecedented happened: the unconscious was put on trial in America.

Ray Belknap was eighteen and James Vance twenty. Two ordinary young men. Ordinary, that is, until 23 December 1985. Together, they listened to some rock music, had a few drinks, smoked some marijuana, and then – for no apparent reason – put shotguns to their own heads. Raymond Belknap was fatally wounded but James Vance survived. He was, however, grossly disfigured and lived for only a few more years. He died in 1988 following a medication overdose.

Why did Ray and James decide to shoot themselves? They might have been typical examples of a disaffected generation, but nothing in their behaviour had signalled a potential suicide risk. How could such a thing happen?

The families of Ray and James began to search for clues. Their search ended among a pile of CDs, where they discovered the 1978 album *Stained Class* by the British rock band Judas Priest – a favourite of Ray and James. It transpired that Judas Priest had placed subliminal messages on *Stained Class*, promoting satanism and kindling suicidal intent. A civil action followed in which the families of Ray and James sought more than $6 million in damages. They argued that listening to *Stained Class* had caused the boys to make a bizarre suicide pact.

The case of Ray and James was so intriguing that the director David Van Taylor chose to make a film about it – the award-winning *Dream Deceivers: The Story behind James Vance vs. Judas Priest*. Moreover, this unique case posed some extraordinary (and disturbing) questions which touched upon matters well beyond the ambit of civil law.

Is it possible to kill someone by enlisting the support of their unconscious? Can the unconscious aid and abet murder? What are the most effective methods of persuading the unconscious to act against the conscious mind? Could subliminal messages replace the assassin's bullet?

Such questions seem more appropriate to the plot of a fictional thriller; yet the case of *Vance* v. *Judas Priest* succeeded in raising these and similar questions in the context of serious scientific debate. Within a few years a special edition of the journal *American Psychologist* was attempting to specify the limits of unconscious influence in a series of invited essays; the first of these – written by psychologists Elizabeth Loftus and Mark Klinger, made particular reference to issues raised during the trial.

So what was the verdict?

The judge presiding over *Vance* v. *Judas Priest* accepted the existence of subliminal messages, but he did not believe that the plaintiffs had produced 'credible scientific evidence' favouring their plea. Judge Jerry Carr Whitehead was not convinced that words instructing listeners to 'do it' could be interpreted by an unconscious region of the mind. Nor was he convinced that such a region could override consciousness. As such, in the eyes of the law, the unconscious is not endowed with sufficient intelligence or influence to precipitate a suicide attempt. In the eyes of the law, the unconscious is relatively stupid – unable to take a hint or respond to the simplest form of instruction. Be that as it may, not everyone agreed then, or agrees now, with these findings.

An altogether different view of the unconscious was taken by Democrats during one of the less creditable episodes of the Gore versus Bush US election campaign in 2000. During a Republican Party television commercial the word 'rats' appeared for one-thirtieth of a second during a sequence in which fragments of the word 'bureaucrats' flickered on and off the screen. The word rats appeared in a larger script, but this could only be detected on careful viewing.

Leading Democrats immediately accused the Republicans of foul play. The producer of the Republican advert, Alex Castellanos, claimed that the very brief appearances of the word rats was 'purely accidental'; however, a commentator familiar with Castellanos and his work stated: 'There is no way anything Alex Castellanos does is an accident'. In an atmosphere of increasing discontent and suspicion, Bush was forced to withdraw the advert.

Opinion polls prior to the 2000 presidential election predicted that the result might be very close. The eventual outcome, of course, confirmed these predictions, when, controversially, the Florida ballots had to be recounted. It is not entirely implausible to suggest that some Republican advisors had come to the conclusion that if the election could not be won by presenting evidence to the conscious mind, then it might be worth presenting evidence to the unconscious mind. With the polls showing equal support for both party leaders, the unconscious emerged as an additional, shadowy constituency.

Although Castellanos' advert produced a flurry of media interest, the stir he created pales into insignificance when compared with the furore generated by the now infamous subliminal scandal of the 1950s. A marketing executive in New Jersey reported that he had superimposed messages on a cinema film reel. These messages appeared so briefly that they escaped conscious detection. Thus, cinemagoers were urged – subliminally – to 'Eat popcorn' or 'Drink Coke'. When the adulterated reel was shown, popcorn sales rose by 58 per cent, and Coke sales by 18 per cent.

The subsequent debate was sufficiently heated and widespread to have repercussions beyond the world of advertising. Very swiftly, academic research into subliminal perception acquired a sinister connotation, and research into preconscious processing was all but abandoned in 'respectable' university departments. Overnight the motives of those with a scientific interest in the unconscious became highly suspect.

Curiously, it may be that research into preconscious processing suffered a serious setback for no good reason. Those involved in the original New Jersey study later announced that the whole thing was nothing more than an elaborate hoax; however, even today it is still unclear what actually happened. Indeed, conspiracy theorists dismiss the disclaimer as a transparent attempt to muddy the waters.

Irrespective of the legitimacy of the original study, the idea that subliminal messages could be used to modify behaviour continued to attract interest in certain quarters. For example, several department stores were rumoured to be using subliminal messages to successfully reduce theft. By embedding messages such as 'If you steal, you'll get caught' in continuous background Muzak, shoplifting rates were allegedly reduced.

The idea of subliminal manipulation (in the form of subliminal advertising or policing) has had an enduring impact on public opinion. Even though the subliminal scandal of the 1950s occurred in the middle of the last century, the concept of subliminal advertising has achieved a remarkable degree of cultural penetration. Most people are familiar with the concept and express appropriate concern when any group – be they a commercial or political organisation – is suspected of employing subliminal messages or instructions. The thought of strangers surreptitiously tampering with our unconscious minds arouses instant anger and resentment.

The issue of subliminal manipulation was considered of sufficient importance to be mentioned in the practice guidelines prepared in conjunction with the 1990 British Broadcasting Act. The Independent Television Commission – who regulate commercial broadcasting – were required to draw up and enforce a code governing standards and practice in TV advertising. It contains the following:

No advertisement may include any technical device which, by using images of very brief duration or by any other means, exploits the possibility of conveying a message to, or otherwise influencing the minds of, members of an audience without their being aware, or fully aware, of what has been done.

A few years after the British Broadcasting Act entered the statutes, the British Psychological Society (BPS) published an official report titled *Subliminal Messages*. The report was compiled largely in response to the sudden appearance of commercially available audio tapes that were supposed to contain improving or healing subliminal messages; however, the content of the report also touched upon more general issues surrounding subliminal manipulation and the unconscious.

The scientific affairs board of the BPS were extremely conservative with respect to their conclusions. Readers were assured that the subliminal advertising scandal in the 1950s was indeed a hoax, that messages of the kind used by Judas Priest on *Stained Class* could not affect listeners, that commercially available tapes promising self-improvement or treatment with subliminal messages did not work, and that it is impossible to learn anything in the absence of awareness (as, for example, during sleep). The report did not question the reality of subliminal perception; however, it suggested that the consequences of subliminal stimulation were at best, fragile and transient. Moreover, readers were reminded that although the brain is capable of processing information outside of awareness, this in itself does not necessarily warrant endorsement of the traditional distinction between conscious and unconscious divisions of mind.

> ... there is no doubt that much mental activity does occur without our being conscious of it, and some of this unconscious mental activity contributes to human learning. However, many psychologists would not agree with the assumption that there are separate conscious minds and unconscious minds. An alternative way of looking at people's mental lives is simply to say that some of the activities of the human brain have consequences that reach consciousness, but others do not. We are often unaware why or how we do things, and we often act for reasons that we fail to remember. But our capacity to absorb and react to complex meaningful messages that are too weak to produce awareness appears to be very limited.

The BPS report is unsensational. Moreover, there is much to be said for restraint in the context of an official scientific report; however, it is possible that the very conservative conclusions reached by the scientific affairs board erred too far on the side of caution. It is indeed true that the evidence for subliminal influence is not universally compelling. Nevertheless, there are many studies in the literature that have yielded positive and impressive results. Therefore, a general statement to the effect that – at best – subliminal stimulation can only ever produce very weak effects may be premature.

With respect to subliminal advertising, conservative commentators maintain that far too much has been made of scandal and hearsay. Moreover, debate over alleged cases of subliminal advertising is entirely sterile, because, quite simply, subliminal advertising doesn't work. But again, such confident assertions might be mistaken.

Polarised opinions are rarely accurate. Claims that subliminal advertising always or never works are very probably both wrong. A much more enlightened approach is to evaluate the effects of subliminal stimulation in a broader context of mediating factors – a point acknowledged by John Kihlstrom in his much-lauded 1987 paper 'The cognitive unconscious'.

Kihlstrom suggests that subliminal instructions might trigger the equivalent of automatic programmes stored in the unconscious. Thus, once triggered, these might influence how consumers think about products, and their subsequent purchasing behaviour. Moreover, subliminal messages may be particularly effective because they obviate the possibility of conscious inhibition. The activation of programmes in the unconscious translates directly into thought or behaviour, unimpeded by self-reflection and careful judgement; however, by the same token, other factors will serve to moderate the potency of subliminal messages.

According to Kihlstrom, subliminal effects are caused by a process described by cognitive psychologists as *priming*. When a subliminal message is presented, corresponding knowledge in the unconscious is activated by the subliminal stimulus. To use a thermal analogy, subliminal presentation of a word such as 'popcorn' will 'warm up' knowledge directly related to 'popcorn' (for example what it looks like, its texture, its taste). Moreover, a little heat might also spread to knowledge that is indirectly related to the concept of popcorn (for example memories of other types of food). The academic literature on spreading activation suggests that priming effects are very short lived. Thus, activation may dissipate as fast as it spreads. This would mean that a subliminal advert might get an individual to think of popcorn for a few moments, but not much more. Moreover, such an individual might not be prompted to think of popcorn for long enough to inspire a subsequent purchase.

Even so, Kihlstrom adds that subliminal priming might serve to amplify (or at least direct) a pre-existing tendency. Thus, if an individual is hungry and he or she receives a subliminal presentation of the word popcorn, it is conceivable that he or she might then begin to desire popcorn; however, if such an individual was not hungry in the first place, subliminal presentation of the word popcorn would probably have little or no effect.

Few people would defend subliminal advertising. It is clearly a cynical, manipulative, and unethical activity; however, after the subliminal scandal of the 1950s, a small number of psychologists were left feeling that some important areas of research had been abandoned. Moreover, the brouhaha surrounding subliminal advertising had produced an intellectual climate in which disinterested discussion could not take place. It seemed that any attempt to exploit the unconscious was automatically considered wrong.

But was this the correct view? Some were concerned that an opportunity was being missed.

If subliminal stimulation of the unconscious was indeed an effective means of producing behavioural change, could the same technique not be used more responsibly – for example to reduce human suffering in the context of mental illness?

The unconscious has been given considerable emphasis in many accounts of mental illness. Although the unconscious is traditionally associated with psychoanalysis, virtually all schools of psychotherapy acknowledge the unconscious in one form or another. Even behaviour therapy, which eschews mentalism in favour of observable behaviour, implicitly recognises unconscious mechanisms in the form of automatic reflexes and conditioned responses. In the context of behaviourism, the conscious, rational mind can do little to moderate phobic anxiety. The acquisition and extinction of fears is governed by processes that operate outside of awareness.

Historically the unconscious has always been linked with the now largely obsolete diagnosis of hysteria; however, when Freud turned his attention to neurotic illness (it was he who coined the term 'anxiety neurosis'), the unconscious was again recruited to explain irrational fears. Many decades later, after the advent of information processing models of mind, the unconscious was used yet again to explain psychotic phenomena such as hallucinations. It was suggested that in problems such as schizophrenia the filter that regulates the passage of information between conscious and unconscious divisions of the mind is damaged, permitting the products of preconscious processing to enter awareness as auditory and visual hallucinations.

The unconscious, however conceptualised, has rarely occupied anything less than a central position in psychopathology; however, there can be little doubt that the link that Freud forged between his particular view of mental illness and the unconscious has proved the strongest.

Psychoanalysis attempts to discover the unconscious roots of psychopathology. Using methods such as free association, dream interpretation, and examination of slips and errors, the analyst gains access to memories and wishes that have been pushed out of awareness. In this sense, the entire endeavour of psychotherapy seeks to 'use' the unconscious.

In psychoanalytic therapy, knowledge of the unconscious confirms theory, informs diagnosis, and facilitates treatment. But therapy can be a protracted process and psychotherapists soon sought to achieve these same ends more swiftly. Freud famously described the interpretation of dreams as the 'royal road' to the unconscious. His followers, however, wanted a freeway.

The first 'artificial' method devised to gain more ready access to unconscious material was the Rorschach test, first published in 1921, which consists of ten symmetrical ink blots (five in black and white and five introducing colour).

The Rorschach test is described by psychologists as a projective test. All projective tests work on the same principle. Typically, an ambiguous stimulus is presented to a patient who then describes his or her understanding of what he or she is seeing. Because the stimulus is intentionally meaningless or ambiguous, the patient's response represents a 'projection'. The patient is seeing something that isn't really there. Thus, the response the patient gives tells the therapist something about the patient's state of mind. Moreover, because the task is relatively innocuous, the unconscious can be 'tricked' into making disclosures. In the same way that the content of the unconscious can be adduced from slips of the tongue and descriptions of dreams, so it is that observations of ambiguous pictures can also be very revealing. Projective tests are a little like magic mirrors. When held up to the mind, they reflect areas of shadow rather than light.

Although the Rorschach ink-blot test is the most famous of all psychological tests, it is also the most maligned. There are many psychologists and psychiatrists who believe that administering the Rorschach test is a largely pointless exercise. Nothing of clinical significance, they argue, can be gained by asking a patient to divine forms in an ink-blot. There may be a kernel of truth in this, yet the Rorschach test continues to attract many advocates. Academic study of the Rorschach test is encouraged by the International Rorschach Society, and an entire journal – the exotic-sounding *Rorschachiana* – is regularly filled with relevant articles. Recently a computerised version of the test was developed, thus securing its presence in the clinics of the twenty-first century.

Curiously, the Rorschach test has now become something of a cultural icon. The image of a symmetrical ink-blot is frequently seen adorning book covers, T-shirts, and CD covers, and it is guaranteed to appear at least once in almost every television documentary on the mind. Even Andy Warhol (a man who knew a thing or two about iconography) produced a series of Rorschach paintings.

Sadly, the Swiss psychiatrist who created this contemporary icon – the enigmatic Hermann Rorschach – did not live to enjoy its cultural popularity. He died only nine months after the first edition was published, at the lamentably young age of thirty-eight.

Another method of provoking the unconscious into revealing its secrets is through subliminal stimulation. This technique involves the presentation of words or images below the awareness threshold, after which patients are instructed to guess what they have just been shown. If, for example, the word 'father' is presented subliminally, and the patient guesses that he or she has been shown a word like 'bully' or 'hatred', it might be supposed that such responses reflect feelings of hostility and, more importantly, hostility that under normal circumstances might not be expressed because of the operation of a defence mechanism. Thus, subliminal provocation can be used to test theory and confirm a psychoanalytic diagnosis.

A good example of how this procedure has been used in practice was published in 1959 by the psychologist H. R. Beech. Beech was interested in the hypothesis that anorexia nervosa is – at least in part – caused by an unconscious confusion between sex objects and food. Anorexic and control subjects received subliminal exposures of various words, after which they were asked to guess what they had been shown. Results confirmed the psychoanalytic hypothesis. Interestingly, those suffering from anorexia thought they had been shown food words after the subliminal presentation of sexual words – an effect which was not observed in healthy controls.

Another psychological test that seeks to tap directly into the unconscious is the defence mechanism test (DMT). The DMT involves the presentation of a picture that shows two figures: a foreground figure (sometimes described as the hero) and a background figure whose expression can be either benign or threatening. The image is presented several times, with each presentation being exposed for an incremental duration. The first presentations are subliminal; however, subjects are still requested to comment on what they have been shown.

Research has demonstrated that when the background figure is threatening, descriptions are more likely to be inaccurate (i.e. elements of the image are distorted or transformed). This phenomenon appears to be consistent with psychoanalytic theory insofar as the perceptual distortions provoked by the presence of a threatening figure may represent the operation of a defence mechanism; anxiety is reduced by transforming the threatening face into a less threatening or more acceptable alternative.

If, as psychoanalytic models suggest, psychiatric problems are associated with the automatic deployment of defences, then the level of distortion reported by DMT respondents will be correlated with the severity of their illness. Moreover, it is implicit that the more defended an individual is, the more out of touch with reality he or she will be. Subsequently the DMT has been employed as an assessment instrument in two quite different settings. First, as a diagnostic test in clinics, and second, as an occupational test for personnel who must demonstrate accurate perception of the environment – particularly under threatening conditions.

After an exhaustive fifteen-year study, the DMT was incorporated into the Norwegian and Swedish airforce selection procedures. The DMT was used specifically to assess pilots applying for flying duties. Follow-up studies have demonstrated that those who perform badly on the DMT (i.e. experience more distortions when the threatening face is in the background) make worse pilots. Such individuals not only exhibit higher levels of absenteeism and psychosomatic illness but, more importantly, show above-average involvement in flying accidents. Similar results have also been found studying deep-sea divers.

Presenting subliminal material to test a theory or to diagnose or assess personality is, of course, controversial; however, such controversy is relatively tame compared with that aroused by the idea of subliminal treatment. First, there is the ethical problem: like subliminal advertising, subliminal treatment is, by definition, non-consensual. Second, there is the problem of credibility: the claim that a single statement or image, exposed for a few thousandths of a second, can have a lasting and beneficial effect on illnesses that are often chronic, complex, and severe seems at first sight to be preposterous. Moreover, the standing of subliminal therapy has not been helped by its cause célèbre, which must rank among the most bizarre discoveries in the history of psychology and neuroscience.

Subliminal therapy, in the form known as subliminal psychodynamic activation (SPA), was developed in the 1960s and 1970s by the late Lloyd Silverman – a psychoanalyst at New York University. Silverman was initially interested in using subliminal stimuli to test – and hopefully confirm – various psychoanalytic ideas about the role of conflict in mental illness. For example, it had been suggested that unconscious aggressive and incestuous urges were important factors in the respective formation of schizophrenic and homosexual 'symptoms'. (Needless to say, contemporary psychologists are inclined to view homosexuality as a preference rather than a psychiatric illness.)

Silverman devised two stimuli which might arouse aggressive or incestuous urges. The first depicted an angry man with bared teeth preparing to stab a woman, accompanied by the words 'Destroy Mother'. The second depicted a nude man and woman in a sexual pose, accompanied by the words 'Fuck Mommy'. Because such violent and libidinal urges are taboo, Silverman's logic was that they would necessarily produce intrapsychic conflict, which would in turn increase symptom severity.

The stimuli that Silverman devised were then shown to schizophrenic and homosexual patients for 4 milliseconds – the idea being to 'stir up' unconscious aggressive and libidinal wishes. Projective tests were administered before and after subliminal stimulation as a measure of psychopathology.

Silverman and his colleagues found that the 'aggression' stimulus activated schizophrenic psychopathology, but had no effect on homosexuals, while conversely the 'incest' stimulus activated homosexual psychopathology but had no effect on patients with schizophrenia.

While Silverman was stirring up intrapsychic conflict, thus making his patients feel worse, he was also thankfully experimenting with more benign uses of subliminal stimulation (i.e. employing the same principles to make his patients feel better).

In the 1960s, there was much interest in psychoanalytic circles concerning early development and the importance of maternal bonding. This coincided with Silverman's own growing interest in mental states associated with fantasies about 'merging with mother' – described by analysts as symbiotic or merging fantasies. Typically, such fantasies evoke feelings of oneness with an idealised nurturing and all-protecting mother.

Silverman speculated that it might be possible to promote feelings of security and contentment in patients by activating unconscious merging fantasies; a goal that he believed could be achieved by subliminal presentations of a key image or phrase. In the end, Silverman chose the words 'Mommy and I are one' – a phrase that he arrived at through chance, rather than design.

The phrase 'Mommy and I are one' was originally spoken by a patient of Silverman's wife – Doris K. Silverman – also an analyst. The patient in question was a troubled young woman whose mother was extremely domineering and sought to control almost every aspect of her daughter's life. In her therapy sessions the patient kept on using the phrase 'Mommy and I are one' to describe the intense nature of their mother–daughter relationship. When Doris Silverman mentioned this to her husband he immediately recognised a phrase that could be used as an economic expression of the symbiotic wish. Thereafter Silverman used the phrase as a subliminal message in many of his subsequent studies, and in the literature it is now known as the 'merging fantasy stimulus'.

Initially Silverman used subliminal exposures of 'Mommy and I are one' with patients suffering from schizophrenia. Careful observations and testing showed a definite improvement in all symptoms, but most notably a reduction in thought disorder. Thus, patients became more coherent and lucid. Inspired by early successes, Silverman and his colleagues then went on the try the merging fantasy stimulus with other patient groups, where subliminal exposures could be used exclusively, or in conjunction with other therapeutic procedures (not all of them based on psychoanalytic theories).

An intriguing example of their work was published in a 1974 edition of the prestigious *Journal of Abnormal Psychology*. Twenty women with insect phobias were allocated to either an experimental or control group. All were given a form of behaviour therapy known as desensitisation. Typically, patients construct a list of situations, ordered to reflect increasing levels of anxiety. Such a list might begin with 'seeing a spider' and end with 'touching a spider'. Patients begin by imagining the first item on their list, but at the same time perform a series of relaxation exercises. This has the effect of inhibiting anxiety. Patients then work their way up the list, pairing relaxation with situations associated with increasing levels of anxiety. Exposure to feared objects or situations can be in imagination, or in real life. Either way, desensitisation sessions of this kind can result in a marked reduction of fear.

In Silverman's study the desensitisation procedure was modified. Instead of receiving relaxation exercises, the experimental group received 4-millisecond exposures of the phrase 'Mommy and I are one'. The control subjects were treated identically, but received 4-millisecond exposures of the theoretically inert phrase 'People are walking'. At the end of treatment, the experimental group showed significantly more improvement than the control group.

Silverman also carried out another series of studies that, while not employing the merging fantasy stimulus, seemed to confirm a cornerstone of psychoanalytic theory – the Oedipal phase of development. This extends from the third to sixth years, during which infant males experience strong hostile feelings (in addition to the more readily recognised benign feelings) towards their fathers, whom they perceive as competitors for their mother's love. Silverman wanted to discover whether activating ideas in the unconscious relevant to Oedipal conflicts could interfere with the execution of a simple task: throwing darts.

A large number of male students were recruited, who then participated in a dart-throwing competition for a financial reward. Some students received subliminal presentations of the phrase 'Beating Dad is wrong', while others received 'Beating Dad is OK'. It was suggested that the idea of winning could be made to resonate with Oedipal implications by subliminal stimulation. Thus, if winning was coloured by the taboo connotation of beating Dad for Mother's forbidden sexual favours, then unconscious inhibitory mechanisms would immediately come into operation so as to impair performance. The opposite, of course, should also be true, where permission will obviate guilt and subsequent inhibition of performance. A control group receiving subliminal exposures of 'People are walking' was also included in Silverman's experimental design.

Results entirely supported psychoanalytic theory. The 'Beating Dad is wrong' group performed worse than controls, while the 'Beating Dad is OK' group performed better than controls.

Since Silverman's pioneering work in the 1960s and 70s, there have been numerous investigations of the therapeutic properties of the merging fantasy stimulus. The most well-researched patient group are individuals suffering from schizophrenia, for whom outcome studies show consistent beneficial effects; however, patients suffering from depression, anxiety (including simple phobias), addictions (such as alcoholism), and eating disorders (including obesity) also show comparable levels of improvement. Finally, in addition to amelioration of symptoms, positive effects of a more general kind have been observed. These include easier self-disclosure, better rapport with others, and increased assertiveness. Used on non-psychiatric subjects, the merging fantasy stimulus has been associated with improved memory and improved academic performance.

In a major review article published in 1990, Richard Hardaway suggested that there is now overwhelming evidence for a slight but consistent effect. Although it is probably the case that the benefits of subliminal exposure to the merging fantasy stimulus have been overestimated in many reports, the fact remains that Silverman undoubtedly discovered a real effect. Hardaway was forced to conclude that 'Future research designed to replicate basic experimental effects is deemed superfluous.' In other words, instead of questioning whether the effect is there or not, the academic community would serve science better by trying to establish how the merging fantasy stimulus works.

This is a sentiment shared by psychologist Joel Weinberger – originally a junior associate of Silverman – who has suggested that the time has come to demystify subliminal psychodynamic activation. To date, advocates have been all too willing to accept Silverman's original explanation for its efficacy or simply assume that some kind of 'magic' takes place. Weinberger suggests that the beneficial effects of subliminal presentations of 'Mommy and I are one' might be mediated by mood. Raised mood is known to facilitate the emergence of a wide range of helpful behaviours (for example increased flex-ibility of thinking, greater willingness to solve problems, and better judge-ment). Thus, Weinberger suggests that although the merging fantasy stimulus is able to elevate mood, this might be the limit of its power. All subsequent benefits are probably derived from raised mood, rather than the activation of unconscious fantasies. Moreover, Weinberger's own empirical work has so far confirmed his hypothesis. Even so, the question of how the merging fantasy stimulus raises mood in the first place remains an interesting and as yet unan-swered question.

Whatever the mechanisms that underlie the efficacy of subliminal psycho-dynamic activation, Silverman's legacy is undoubtedly challenging and thought-provoking; however, perhaps these qualities were also characteristic of Silverman himself. Not only did Silverman choose to begin his research programme in a decade when research into preconscious processing was dis-couraged, but he then chose to continue this programme into a decade when psychoanalytic ideas were routinely vilified by the academic establishment. He seemed to invite criticism, but when it arrived he had sufficient elan to respond by publishing an article in *American Psychologist* with the mischievous title 'Psychoanalytic theory: "the reports of my death are greatly exagger-ated"'.

When Silverman submitted his early work on subliminal psychodynamic activation to the *Journal of Abnormal Psychology* it was swiftly rejected. Silver-man was told bluntly that his results were simply unbelievable and therefore could not be published. Whereas most academics in Silverman's position would have just given up at this point, Silverman was more than happy to contest the decision. He took his case to the American Psychological Associ-ation, and after an independent investigation the editorial decision to reject his article was overruled. Silverman's work was deemed to be of an extremely high standard and obviously fit for publication.

Clearly, Silverman relished controversy. He never dismissed ideas – however outlandish – and believed that any issue could be resolved by designing and conducting the right experiment. Indeed, he was such a com-mitted empiricist that his wife had the epithet 'It's an empirical question' carved on his tombstone. Silverman urged his students to question virtually everything, and inspired generations of researchers (who still remember him with great affection). As a teacher, he is reputed to have been second to none. It would be a mistake to label Silverman as a peripheral figure or a maverick. He ranks among only a handful of psychoanalysts interested in testing psy-choanalytic ideas in a laboratory setting, and in many respects his work rep-resents the most successful and exciting of its kind.

Given the importance of unconscious processes in psychoanalytic theory, it is not surprising that the psychoanalytic community has expressed most interest in the therapeutic possibilities of subliminal stimulation; however, the literature on subliminal treatment does contain one conspicuous and notable exception – a report that describes a successful behavioural treatment administered outside awareness.

The behavioural treatment for anxiety involves encouraging patients to confront their fears. Sometimes confrontation is paired with relaxation exercises (as in desensitisation) and sometimes not. Opinion differs between behaviour therapists regarding the value of relaxation exercises performed during episodes of confrontation; however, with or without relaxation, initial exposure to feared objects or situations always produces some anxiety in the patient.

It is possible that behaviour therapy works by exploiting a property of the nervous system called habituation – a progressive tendency to stop responding to repeated presentations of the same stimulus. Habituation phenomena are happening all the time; for example we habituate to background noise – such as traffic or a ticking clock: after a time, we simply stop hearing them. If a behaviour therapist can get his or her patient to remain in a feared situation for a sufficient length of time (usually about half an hour), then the parts of the nervous system producing the symptoms of anxiety will become less active. The more a patient is exposed to a feared object or situation, the more he or she will experience a reduction in anxiety. The difficulty, of course, is getting patients to stay in an anxiety-provoking situation long enough for beneficial effects to materialise. Many patients fail to engage in behaviour therapy because they find the whole procedure too traumatic. They want to leave the feared situation before habituation has occurred.

In 1978 the psychiatrist Peter Tyrer and his colleagues published an article in *The Lancet* titled 'Treatment of agoraphobia by subliminal and supraliminal exposure to phobic cine film'. In order to make behaviour therapy more tolerable, Tyrer and his colleagues attempted to produce habituation in anxious patients by exposing them to images of fearful situations presented below the awareness threshold.

Thirty agoraphobic women were recruited for the experiment, all of whom had proved resistant to conventional treatment. A supraliminal group watched a film showing scenes likely to trigger fear in someone suffering from agoraphobia (i.e. wide open spaces). A second, subliminal group, were shown exactly the same film, but at an intensity too weak for conscious perception. Finally, a control group watched a blank screen which alternated with a sequence of 'a potter working at his wheel'.

The results were extremely interesting. Both the subliminal and supraliminal groups improved more than the control group, becoming less anxious and agoraphobic; however, the subliminal group showed a modest advantage, being a little less depressed than the supraliminal group.

Patients in the subliminal group were at a loss to explain their improvement. This is understandable, as their experience of treatment was simply watching a blank screen. Subsequently, several patients in the subliminal group concluded that the procedure was a subtle means of letting them know that if they wanted to get better then they had better help themselves. This represents yet another example of the conscious mind generating a plausible story, to account for behaviour shaped by events and processes occurring outside of awareness.

Clinical psychologists, psychoanalysts, and psychiatrists have barely begun to explore the therapeutic potential of channelling information directly into the unconscious. Even so, initial results are highly promising. Moreover, given that the demand for psychiatric care far exceeds the supply of relevant services, the appeal of treatment methods that can achieve small but consistent health benefits (sometimes in under a second) is patently obvious; however, irrespective of practical advantages, subliminal interventions will always be dogged by a moral question: Is it acceptable to administer a treatment for which informed consent cannot be given? Until society reaches some consensus on this issue, subliminal treatment will have trouble escaping the long shadow cast by subliminal advertising. (Interestingly, Silverman was frequently approached by advertising executives eager to explore the commercial possibilities of the symbiotic merging fantasy stimulus as a retail tool.) In addition, there is another perhaps even more significant factor that has limited the widespread use of subliminal treatments – professional incredulity. In spite of the evidence, most psychiatrists and psychologists simply don't believe the positive results that have been reported in the academic literature. Thus, the potential of subliminal therapy – be it modest or revolutionary – will in all probability never be fully realised.

For most people, nightmares are of little significance. Beyond infancy and childhood the distress caused by bad dreams is nothing more than a minor inconvenience. Severe and frequent nightmares, however, are a very different matter. They can be so intense and disturbing that the afflicted individual is completely unable to function. Nightmares of this kind are recognised as a form of psychiatric illness, and guidelines for the diagnosis of *nightmare disorder* have been established by both the American Psychiatric Association and the World Health Organization. Nightmares also frequently occur in the context of other psychiatric problems – in particular, anxiety and post-traumatic stress.

Conventional treatments for nightmare disorder are pharmacological (i.e. using drugs) or behavioural. The behavioural treatment of nightmares is somewhat restricted insofar as it is only applicable to recurring nightmares (i.e. a single nightmare that returns again and again). Treatment involves imaginary rehearsal of the nightmare while awake – but with a modified and preferably happy ending. Eventually this modified version of the nightmare replaces the original during sleep.

A potentially superior treatment for nightmares – the acquisition of lucid dreaming skills – has been available for over a century; however, mainstream practitioners have been reluctant to exploit it.

Hervey de Saint-Denis, the man who discovered lucid dreaming in the late nineteenth century (see Chapter 1), was quick to explore its therapeutic possibilities. Indeed, he managed to cure himself of his own recurring nightmare – which involved being pursued through an endless series of rooms by horrific monsters.

> I stared at my principal assailant. He bore some resemblance to one of those bristling and grimacing demons which are sculpted on cathedral porches. Academic curiosity soon overcame all my other emotions. I saw the fantastic monster halt a few paces from me, hissing and leaping about. Once I had mastered my fear his actions appeared merely burlesque. I noticed the claws on one of his hands, or paws, I should say ... the result of concentrating my attention on his figure was that all his acolytes vanished, as if by magic. Soon, the leading monster also began to slow down, lose precision, and take on a downy appearance.

Typically, training in lucid dreaming involves cultivating a special type of self-awareness that carries over into the dream world. At its simplest, this can be achieved by frequently asking the question 'Am I dreaming?' while awake. Eventually, the very same question is posed while dreaming. The ensuing awareness of being in a dream is generally sufficient to enable the dreamer to alter its course.

Since the time of Hervey de Saint-Denis, only a small group of psychologists and psychiatrists have expressed an interest in lucid dreaming. Even so, the results of these exploratory investigations have yielded some interesting results.

It would appear that the feeling of being in control (i.e. being able to influence the course of a dream) acquired during lucid dreaming often engenders a similar mental state in real life. Thus, patients report being more confident and assertive. There are even descriptions of benefits of a less-tangible – even transcendental – nature. For example, in *The Sun and the Shadow* (1987), the psychotherapist Kenneth Kelzer wrote of one of his own lucid dreams:

> In this dream I experienced a lucidity that was so vastly different and beyond the range of anything I had previously encountered. At this point I prefer to apply the concept of the spectrum of consciousness to the lucid dream and assert that within the lucid state a person may have access to a spectrum or range of psychic energy that is so vast, so broad and so unique as to defy classification.

Kelzer's descriptions of the dream world are reminiscent of Jung's. One is reminded of Jung's descent into his own unconscious, his encounters with archetypes, and his return as the custodian of arcane knowledge. Although such excursions into the unconscious make compelling reading, most therapeutic applications of lucid dreaming do not rely on contact with the numinous. Indeed, most applications of lucid dreaming rely on more straightforward mechanisms to explain beneficial change, such as the release of emotion or the rehearsal of a skill.

Lucid dreams can be liberating or cathartic. In the dream world, it is possible to engage in behaviours that would be considered unacceptable in real life. Thus, the lucid dream can become a kind of 'natural' virtual-reality game. The unconscious is transformed into a nocturnal playground where frustrated emotions can be vented, forbidden desires satisfied, and physical limitations ignored. The latter may be of particular benefit to the elderly or those suffering from physical disabilities.

With respect to the acquisition of skills, some studies have shown that sportsmen exhibit improved performance after rehearsing complicated manoeuvres in their lucid dreams.

A further application of lucid dreams (so far overlooked) is as a possible environment for the practice of behaviour-therapy exercises. Perhaps patients wary of confronting feared objects and situations in real life could be encouraged to do so in their dreams. Thus, exposure exercises are undertaken in the 'unconscious' rather than in the clinic. Could such a procedure be as effective as conventional exposure? Or might it even produce superior results?

A less dramatic although by no means less interesting method of modifying dreams is the *Poetzl effect*. This unusual phenomenon was first reported by Otto Poetzl in 1917, following observations of patients who had sustained damage to the visual areas of the brain. Poetzl discovered that stimuli failing to register in awareness would subsequently enter awareness by some other (often indirect) process. For example, one of Poetzl's patients was unable to perceive an asparagus fern protruding from a bunch of roses; however, subsequently the same patient showed highly selective attention for a green tiepin which resembled the fern on the collar insignia of a uniform.

Poetzl suggested that in some neurological patients, the mechanisms that suppress information – particularly information that has been registered in the brain but not represented in awareness – break down. Subsequently suppressed information enters awareness, although not necessarily in its original form. Perceptual biases might cause a related object to stand out in the perceptual field (as in the fern and tiepin example), or the original information may enter awareness having undergone some order of transformation (preserving, perhaps, a symbolic relationship).

Poetzl then argued that the healthy sleeping brain might recreate the very same effect. Thus, information that has been registered during the day – but not represented in consciousness – will then appear in dreams. Moreover, a corollary of Poetzl's argument is that subliminal presentations will also re-emerge from the processing system in the narrative of dreams – albeit in a symbolic or distorted form.

Although the reasoning that guided Poetzl from clinical observations to the laboratory was somewhat suspect, the research he undertook was very encouraging. In his first experiment, Poetzl showed subjects a colour slide of the Temple at Thebes for one hundredth of a second – which is long enough to see, but only fleetingly. Subjects then described and drew what they had seen. Following this laboratory task, subjects were instructed to keep a detailed record of their dreams, to draw sketches of what they saw in their dreams, and, finally, to examine the original picture and compare it with their records and drawings. Interestingly, only those features of the picture unreported by subjects after the initial showing appeared again in subsequent dreams.

Much work on the Poetzl effect has been undertaken since 1917, and the phenomenon seems to be fairly robust. Indeed, the Poetzl effect reliably survives different modes of presentation. If stimuli are presented subliminally, or if stimuli are presented supraliminally (but are unattended), links can be made between original presentations and the later entry of related material in awareness. Although the return of information in dreams is the most well-researched example of the Poetzl effect, many other methods have been used to catch these extraordinary rebounds from the unconscious, for example, fantasy drawings, doodles, responses to the Rorschach test and word association.

What then, are the practical uses of the Poetzl effect?

Firstly, there is some evidence to suggest that Poetzl-type phenomena can be employed diagnostically. The unattended elements of pictorial presentations that appear in dreams may indicate areas of psychopathology that are ordinarily inaccessible to consciousness. Secondly, and more speculatively, the Poetzl effect might be used to exercise a benign influence on the content of dreams. Again, this would be most relevant for individuals suffering from chronic nightmares. Subliminal presentations of comforting images might possibly 'crowd out' nightmare images. Such a treatment would be particularly useful for very young children, for whom drugs or behaviour therapy might not be an option. Interestingly, the Poetzl effect has been reported in the context of the symbiotic merging fantasy stimulus. Some heroin addicts who received subliminal exposures of 'Mommy and I are one' later reported dreams in which a mother figure rescued a child.

Most people have had the experience of struggling to solve a problem, having to give up on account of its difficulty, only to have the solution miraculously pop into awareness at a later stage. This phenomenon is all the more remarkable because usually no effort to solve the problem is made between giving up and getting the answer. So well recognised is this effect that people grappling with seemingly intractable problems are routinely advised to 'sleep on it'.

Psychologists have called the period that extends from 'giving up' to 'getting the answer' the incubation period. Although conscious thought is abandoned, it is assumed that important unconscious or preconscious processes are taking place. These unconscious processes that assist in the solution of problems sometimes become visible in dreams. An image or symbol appears which somehow encapsulates the essence of a solution.

Descartes famously resolved to develop his own approach to philosophical enquiry after a complex dream involving whirlwinds, cryptic books, and mysterious figures; the physicist Niels Bohr gained insight into the structural properties of the atom when he had a dream about a racecourse in which the horses ran in lanes corresponding with the orbit of electrons; and the chemist Friedrich Kekulé solved the structure of the benzene ring when, while dozing, he had a vision of hydrogen and oxygen atoms joining together in a circle.

However, it is artists, rather than scientists, who seem to have drawn the greatest inspiration from dreams. The romantics positively cultivated a rich dream life in order to liberate fully formed works of art from the unconscious. Coleridge's 'Khubla Khan' remains the seminal example, but there are many more. The poet and artist William Blake claimed to have learned a new engraving technique (relief etching) from his dead brother who appeared to him in a vision or 'dream', and the composer Tartini first heard the music of his *Devil's Trill* sonata played by the devil in a dream. The music of Stravinsky's *The Rite of Spring* – perhaps the most influential composition of the twentieth century – was also first heard in a dream.

Mozart, on the other hand, did not need to be in a dream to tap his uniquely productive unconscious. Inspection of Mozart's original manuscripts shows that his compositions went through very few drafts and required comparatively few corrections. Moreover, it is well known that Mozart could compose while he seemed to be fully engaged in other activities. A game of billiards didn't stop him from composing masterpieces. Even though he was concentrating on a shot, the process of composition proceeded without effort. It is also clear that Mozart was able to conceive compositions 'whole'. Complete works, that would unravel over extended periods of time in performance, seemed simply to appear in his mind, polished to perfection in a single moment.

Such extraordinary feats strongly suggest the presence of a hidden intelligence, continuously labouring below the threshold of awareness. It seems implausible to suggest that Mozart's endlessly inventive music was the result of compositional skills that had become automated through practice; however, the potential of insensible neural reflexes shouldn't be underestimated.

It is well known, for example, that the Russian composer Dmitri Shostakovich could hear music if he tilted his head to one side. This bizarre phenomenon was the direct result of an injury. During the First World War a splinter of metal from a shell became embedded in his brain. Presumably, when Shostakovich tilted his head, the splinter moved and stimulated the auditory apparatus. This resulted in the automatic production of perfectly coherent music. Allegedly, some of the music that Shostakovich heard in this way was so impressive he later incorporated it into his compositions.

Whether works of art produced with little or no conscious effort are the result of a hidden intelligence or a set of neural reflexes is a debate of hardly any concern to creative artists. They are much more interested in the more pressing issue of exploiting the unconscious.

In many respects, the methods of exploiting the unconscious favoured by the romantics are linked by self-annihilation. Opium, merging with nature, pursuing extremes of emotion (even to the point of welcoming insanity), are all ways of dissipating or weakening the everyday, conscious self. This is of considerable interest insofar as contemporary research has shown that self-awareness interferes with task performance – particularly if the task in question has been automated through practice. Thus, in many respects, it makes perfect sense to eliminate conscious control of the creative process. The mechanisms that govern creativity will work more efficiently if consciousness has been dissolved – albeit temporarily.

The tradition of self-annihilation did not end with the romantics. Indeed, many twentieth-century artists, particularly those who broke new ground, aspired to a state of non-being while working. William Burroughs, for example, is supposed to have written *The Naked Lunch* (1959) in a state so dissolute that he had no recollection of having written any of it. Moreover, the painter Francis Bacon worked long hours, ceaselessly modifying images, in an attempt to lose his ego and facilitate the transition of unconscious material directly on to the canvas.

We live in a culture which deifies artists. Critics and academics pore over biographical details in an effort to establish links between the artist's life and work, but the entire enterprise might be misconceived. Examining an artist's ego will more than likely only reveal inhibitory factors. To understand the dynamics of creativity requires access to an altogether different biography, the secret biography of the unconscious.

Ironically, a potentially beneficial use of the unconscious is featured in Aldous Huxley's dystopian novel *Brave New World*. In Huxley's chilling vision of the future, children are educated (or rather indoctrinated) in their sleep, using a technique called *hypnopaedia*. The advantages of being able to deliver education during sleep are obvious. Information can be acquired without any effort and waking hours can be used for more leisurely pursuits. Unfortunately, until recently, sleep-learning had little, if any, scientific credibility. Over the past twenty-five years, however, evidence for the efficacy of sleep-learning has been mounting. Moreover, much of this evidence has sprung from an unexpected source – the literature on surgical anaesthesia.

For as long as general anaesthesia has been in use, surgeons have been intrigued by patients who, although entirely unconscious during the course of an operation, wake to report certain experiences. The most detailed and expansive of these are the now well-known out-of-body and near-death experiences. Typically, patients report leaving the body, being able to observe the surgeons operating, and journeying through a long tunnel to a realm of light. At the more pedestrian end of the spectrum are reports of theatre staff conversations that took place while the patient was supposed to be asleep. Although not as impressive as near-death experiences, such reports are nevertheless extremely puzzling. They seem to suggest that the unconscious brain can listen, learn, and remember.

Inspired by these anecdotal reports, a small group of researchers began to

test the brain's ability to learn under general anaesthesia. An early and memorable study conducted in the mid 1960s, involved the staging of a mock
crisis during surgery. It was found that four out of ten patients could recall
the crisis under hypnosis, and another four could remember that there was
some sort of problem but could not remember the specific details.

From the 1980s investigations into learning during general anaesthesia
became less dramatic but more controlled. After receiving a general anaesthetic, patients were 'played' informative tapes. Prior to surgery, the experimental group answered 37 per cent of general knowledge questions
correctly. After surgery, their hit rate rose to 62 per cent.

More recently psychologists have focused on more subtle procedures to
detect learning during anaesthesia. In a typical contemporary study, a recording of a word list is played to patients while they are undergoing surgery.
After waking, they are given a special kind of memory test – known as an
implicit memory test – to see if any of the words were registered in the
unconscious brain.

The term 'implicit memory' is used by psychologists to describe memory
for knowledge which is not associated with a particular learning experience.
We have already encountered this concept in the context of patients suffering from anterograde amnesia (see Chapter 7), who can learn new skills and
facts but cannot remember acquiring this information. They feel like they are
guessing the answers to questions, but discover that they are guessing right.
Implicit memory can be thought of as unconscious memory. Therefore, tests
of implicit memory are designed to detect unconscious learning.

A memory test of this kind might involve presentation of the word *javelin*
to a patient during surgery. After waking, the patient might then be shown
the letters *j* and *a* and instructed to think of a word beginning with these two
letters. If learning has occurred, then the patient is more likely to reply
'javelin' than 'jam' or 'January'.

There is now quite a large body of work on the subject of implicit memory
for words presented during general anaesthesia. Approximately half of these
studies have produced positive results. That is, patients completed unfinished letter stems using words that they had heard while asleep more frequently than would be expected by chance. Although only half of the studies
have produced positive results, this is by no means insignificant. The fact that
sleep-learning occurs at all is in direct contradiction to received scientific
wisdom.

Even so, a few caveats are in order. Recent findings suggest that memory
for information presented during surgery is somewhat short lived. Thus, positive results are more likely to be achieved if memory tests are administered

immediately after surgery. The longer the interval between surgery and the administration of implicit memory tests, the fewer the words remembered. In addition, there seems to be a relationship between memory effects and depth of anaesthesia. Shallow anaesthesia produces better implicit memory for words than deep anaesthesia; thus, it would appear that the closer the sleeping brain is to consciousness, the more receptive it is to learning.

The somewhat patchy and inconsistent results obtained from investigations of sleep-learning have been explained with reference to a number of factors. Obviously, fluctuations in depth of anaesthesia ranks high among them; however, in addition, it has also been suggested that hormone levels have a significant impact on the brain's capacity to register or record information.

From a subjective point of view, the brain is experienced as being 'on' or 'off'. Even in dreams, we experience the brain as being switched on. Although dreams might well represent activity in the unconscious, we, as observers or actors in the dream, are definitely not unconscious. When we reflect on our dreams, we endorse an implicit assumption that the brain is operational.

The sleep-learning literature seems to suggest that our subjective experience of the brain being 'on' or 'off' is misleading. In reality, between the two theoretical extremes of coma and hypervigilence might be a graded continuum of functioning. Moreover, some of these levels of functioning proceed efficiently in our absence. Even though we are switched off, our brains are still switched on. If this were not the case, then experimental investigations of learning under general anaesthesia would not have produced a single positive result.

In this sense, it would be true to say that the unconscious never sleeps.

10

The third blow

Freud's celebrated assertion that – after Copernicus and Darwin – he had delivered the third and most wounding blow to a narcissistic human race appears in his *Introductory Lectures on Psychoanalysis*. According to Freud, 'the ego is not even master of its own house, but must content itself with scanty information of what is going on unconsciously in the mind'.

It isn't necessary to be a disciple of Freud, or a devotee of psychoanalysis, to subscribe to this view. The pith of Freud's argument can be drawn out from the husk of psychoanalytic theory without loss of integrity and coherence: human thought and behaviour are determined largely by unconscious processes. We obey orders that are issued from below the threshold of awareness, and we obey like automata.

Freud's insight was not new. The idea that complex behaviours might be produced by insensible reflexes in the nervous system had been proposed before Freud's birth. The distinctive feature of Freud's thinking is not that he recognised unconscious determinants of behaviour, but that he afforded this insight such monumental importance. He ranked it with the two greatest discoveries in the history of science: the heliocentric universe and evolution. A bold and seemingly outrageous claim, but also a bleak and pessimistic one.

If all human behaviour is ultimately determined by unconscious processes, to what extent can human beings really be described as free? To what extent can human beings exercise rational judgements? To what extent are the identities we cherish anything more than epiphenomena; a superficial skein of thoughts and emotions, serving only to obscure the more significant workings of the unconscious?

The conclusions that follow from Freud's understanding of the mind are depressing. He was never sanguine about the human condition, and famously described the 'common unhappiness' of everyday existence; however, in the *Introductory Lectures on Psychoanalysis* Freud's pessimism takes a more epic form. His third blow is aimed not only at the individual but at the species. Freud's words are coloured with emotions that are close to contempt – or perhaps even disgust. He writes of man's 'naive-self love', 'ineradicable animal nature', and 'megalomania'.

Why? What had made Freud look so closely, and so critically, at the human race? What had compelled him to take a hammer to human pride?

The *Introductory Lectures on Psychoanalysis* were originally planned as talks, and were delivered in the course of two winter programmes between 1915 and 1917. Freud's audience consisted of assorted academics from different departments at the University of Vienna.

Freud was nearly sixty when he began the first of his lecture programmes. This in itself may have been enough to lower his spirits. At several junctures in his life, certain ages had come to be associated with his own death. He had expected to die at fifty-one – but didn't. Subsequently, he assumed he would die aged sixty-one. So, he was probably somewhat cowed by the prospect of turning sixty.

About this time Freud had begun to complain a great deal about feelings of tiredness and fatigue. Although his followers praised his intellectual energy, Freud was inclined to remind them that, as far as he was concerned, he was feeling his age. Establishing psychoanalysis had been an exhausting endeavour. He was not just an old man, but a very weary old man.

And of course, as he mulled over the *Introductory Lectures on Psychoanalysis*, he did so against the historic backdrop of a world at war. A war made more personal and worrying for Freud when his two sons, Ernst and Martin, were posted at the front.

Yet, apart from one brief and exceptional passage, there are no direct references to the war in the *Introductory Lectures*. He merely wrote about psychoanalysis. Although at the outbreak of war Freud had showed considerable interest in its progress, by 1917 his only wish was that the war would end. In Vienna, food and fuel had become scarce commodities. Most evenings he would retire to his study – a hungry old man – where he would record his observations of human nature with cold benumbed fingers. Curiously, he took to collecting jokes about the war in an effort to cheer himself up.

In spite of the prevailing atmosphere of gloom and despondency there was a ray of hope for Freud – it was rumoured he might be awarded the Nobel Prize. But even this failed to materialise. On 25 April he noted, rather laconically: 'No Nobel Prize, 1917'.

So these were the circumstances in which Freud formulated and delivered the third blow. Old, cold, hungry, and preoccupied with his own mortality, worried about his sons, and anxious for the war to end. Clearly, ideal conditions for his constitutional pessimism to thrive.

In Lecture 9, which deals with the subject of censorship in dreams, Freud's feelings about the war and human nature suddenly intrude. A paragraph – only loosely connected to the preceding line of argument – becomes an unexpected channel for Freud's spleen.

And now turn your eyes away from individuals and consider the Great War which is still laying Europe waste. Think of the vast amount of brutality, cruelty and lies which are able to spread over the civilised world. Do you really believe that a handful of ambitious and deluding men without conscience could have succeeded in unleashing all these evil spirits if their millions of followers did not share their guilt?

Here, then, is the truth about human nature. According to Freud, it is senseless blaming politicians for the folly of war. The responsibility for such widespread brutality and cruelty is shared by a complicit humanity. The 'evil spirits' that Freud describes are in all of us.

Freud invokes the concept of guilt, but only for dramatic effect. To be guilty, an individual must make choices, and ultimately Freud does not believe that human beings are capable of making choices – at least not in any absolute sense. If behaviour is largely determined by unconscious processes, then the self is disenfranchised. We subscribe to a collective delusion of agency, but in reality the strings of the human marionette are pulled and released by powers that operate beyond the scope of awareness. Observing world affairs between 1915 and 1917, Freud was inclined to view these powers as demonic.

Through the first half of the twentieth century public perception of Freud underwent a radical transformation: the Viennese doctor with eccentric ideas about sex and dreams slowly evolved into a cultural icon. The latter half of the century reversed this process. Since the 1950s, Freud has been routinely defamed, rebuked, and slandered; however, now that both groups, evangelists and iconoclasts, have exhausted their respective supplies of zeal and bile, the prevailing, somewhat cooler atmosphere is facilitating a more accurate assessment of Freud's standing.

In the early years of the last century science was animated by three big ideas: the gene, quantum physics, and the unconscious. It is commonly believed that genetics and quantum physics realised their early promise, whereas the unconscious led psychology and neuroscience down a conceptual and theoretical cul-de-sac.

It is certainly the case that psychoanalysis did not live up to initial expectations. The claims made for psychoanalysis as a treatment were always inflated; it wasn't original, nor did it prove to be a panacea. As a general framework for understanding the mind and human behaviour, psychoanalysis has much to offer, but it also has profound weaknesses. In the 1970s and 1980s, the psychologist Hans Eysenck and the philosopher Adolf Grunbaum published penetrating critiques of psychoanalysis, revealing flaws and problems of a very fundamental nature – although many equally devastating critiques were also produced by those working within the psychoanalytic movement itself. Even so, Freud made many assertions that have stood the test of time, and among these we must now count his assertion that the determinants of behaviour are mostly unconscious – the nub of his 'third blow'.

Unfortunately, psychoanalysis and the unconscious have been tarred with the same brush. Those who rejected psychoanalysis also rejected the unconscious as a matter of course; however, not only is it possible to reject psychoanalysis and at the same time accept the importance of the unconscious, it is also possible to reject psychoanalysis and accept some of Freud's conclusions about the importance of the unconscious.

Freud was insistent that the most fundamental questions about human mental life could only be answered with reference to a stratum of mind unavailable to conscious inspection. Curiously, most contemporary philosophers, neuroscientists, and psychologists, have reached exactly the same conclusion. The unconscious is no longer a conceptual and theoretical cul-de-sac. The unconscious is an open road, extending to a horizon beyond which lies the future of brain sciences.

For centuries, the holy grail of philosophy and neuroscience has been an explanation of consciousness. Arguably, it is the most fundamental of all questions. How does the loose porridge of neurones and neurotransmitters filling the human cranium produce the miracle of awareness? It seems impossible that experiences such as seeing the colour blue, falling in love, or enjoying the fragrance of a rose are all ultimately generated by a clod of brain matter. Consciousness seems to be qualitatively different from everything else that exists in the known universe: a divine flame. This was certainly the view of the philosopher Descartes.

Much of what Descartes had to say on this subject is written in his lyrically titled *Treatise on the Passions of the Soul*, which was published in 1649. According to Descartes, there are two kinds of substance that exist in the universe. The material substances (res extensa), that make up the world and the brain, and an immaterial substance (res cogita), from which the mind is formed. He wrote: 'this ego, this mind, this soul, by which I am what I am, is entirely distinct from the body'; however, in the Cartesian system, the mind is somehow linked to the brain by the pineal gland, and is thus able to influence the rest of the body through it. Descartes' account of the mind and body is described as 'dualistic' because it requires the existence of two quite different substances – 'mind stuff' and ordinary matter. Unfortunately, there is no evidence whatsoever for the existence of 'mind stuff'. Regardless of how closely the contents of the skull are examined, nothing other than material substance appears to be present. Thus, investigators are left with what the philosopher Gilbert Ryle derisively called 'a ghost in the machine'.

There is nothing ghostly or speculative about a brain. Yet, belief in a gossamer mind, snagged and trapped in the biological machinery, remains an enduring and remarkably tenacious image. Even today, in spite of prodigious advances in brain sciences, there are many well-informed commentators who are prepared to argue that consciousness is completely beyond the reach of human enquiry.

A common defence of this position is that the brain is too insubstantial an organ to account for what a human being is capable of experiencing. There just doesn't seem to be enough of it. But exaggerating our own mystery might be yet another example of what Freud would automatically describe as human narcissism; another attempt to preserve an increasingly tired and shoddy claim on divine provenance. There are 10 to the power of 10 plus neurones in the brain; and, considerably more synapses (junctions involved in the transmission of chemical signals). To count the number of connections between cells in the outer layer of the brain alone – at the rate of one a second – would take approximately thirty-two million years. That is, roughly the same time as it has taken the very early apes to evolve into human beings. There are, in fact, more neurones in a single brain than there are stars in the known universe. Why shouldn't such an object be capable of generating consciousness? However, even if it is provisionally granted that the brain is sufficiently complex to produce consciousness, the magic that turns biological complexity into a person remains opaque.

One of the most widely cited biological explanations of consciousness is the so-called *homunculus solution*. This is the suggestion that the brain contains an area designate where images and impressions are interpreted and bound together. Thus, consciousness is manufactured in a precise location – albeit one difficult to determine.

This idea is known as the homunculus solution because it implies a central executive: a little man, reduced in size, occupying a chamber in the brain from where he performs his singular and very demanding task.

Unfortunately, the homunculus solution has two very fundamental problems. Firstly, in spite of considerable efforts to find the homunculus's hideaway, it seems that there is no specific area in the brain which monitors, combines, and interprets incoming sensory information; secondly, the homunculus solution immediately recreates the initial problem: how does the homunculus achieve consciousness? Another, even smaller homunculus? The ultimate solution is never provided. Instead, it is postponed indefinitely by an infinite regress of increasingly diminutive homunculi.

The homunculus solution highlights the problems associated with trying to explain something by simply describing it in a different way. Suggesting that consciousness is explained by processes that operate in the 'consciousness producing area' of the brain does little to advance our knowledge. In order to understand how a flame works, it is necessary to understand how fuel is ignited in the presence of oxygen. Observing that a flame flickers and emits heat is merely a description – one of many possible descriptions. For centuries, philosophers and scientists have attempted to understand consciousness by scrutinising and describing the flame of awareness. Paradoxically, the answer to the riddle of consciousness might not be in the light, but in the darkness.

There is now widespread agreement that consciousness is an emergent property. Scanning studies show that brain activity is distributed among numerous specialised sub-systems. None of these is conscious. None of these can think, talk or feel; however, working together, the ensemble seems to generate something greater than the sum of its parts. Given the complexity of the brain, it might be that consciousness spontaneously arises once the nervous system evolves beyond a critical threshold of information-processing density. In all probability, understanding consciousness will be better served by examining the unconscious processes that generate consciousness, rather than consciousness itself.

The concluding section of the philosopher Daniel Dennett's seminal work *Consciousness Explained*, published in 1991, contains a telling passage:

> Only a theory that explained conscious events in terms of unconscious events could explain consciousness at all. If your model of how pain is a product of brain activity still has a box in it labelled 'pain', you haven't yet begun to explain what pain is, and if your model of consciousness carries along nicely until the magic moment when you have to say 'then a miracle occurs' you haven't begun to explain what consciousness is.

Virtually every major advance in neuroscience since 1991 has added weight to this same conclusion. The answer to the problem of consciousness will be found below the limen rather than above it.

The unconscious and unconscious processes are now routinely invoked by contemporary philosophers and neuroscientists. In his 1999 work, *The Feeling of What Happens: Body and Emotion in the Making of Consciousness* Antonio Damasio underscores the enormous amount of information that is stored and processed unconsciously in the human brain. In his exhaustive survey he includes all the fully formed images to which we do not attend; all the neural patterns that never become images; all the response dispositions that were acquired through experience, remain dormant, and might never become an explicit neural pattern; all the preconscious modification of such dispositions (that might never become explicitly known); and 'all the hidden wisdom and know-how that nature embodied in innate, homeostatic dispositions' (i.e. the self-regulatory systems in the brain).

For Damasio, the concept of the unconscious is entirely integrated into his model of the mind–body relationship. The unconscious has a biological taproot in the body itself. The body regulates its systems without awareness, but its choices are 'intelligent'. Appropriate chemicals are automatically released to help us digest our food or run away from danger. The immune system learns to recognise invading organisms and can mobilise specialised cellular forces to deal with subsequent threats. The body 'observes', 'learns', and 'understands'. But it does so in ignorance. It performs these miracles unknowingly.

In the 1960s and 1970s, 'respectable' brain scientists would have eschewed concepts such as unconscious emotions and unconscious selves. Yet, in Damasio's most recent theoretical formulation, both unconscious emotions and the idea of an unconscious proto-self are fully embraced.

Of course, the unconscious of Damasio and his neuroscientist peer group is not the unconscious of Freud and Jung. It is an unconscious that has much more in common with Laycock's 'reflex function of the brain', T. H. Huxley's automatic processes, and Kihlstrom's cognitive unconscious. It is an enabling unconscious. Even so, the emphasis that contemporary neuroscientists give to unconscious processes is fundamentally in accord with Freud. They, like him, are prepared to accept that events occurring outside of awareness are the principal determinants of human behaviour. In many respects, this basic consensus is far more significant than all the discrepancies that bring Freudian ideas into conflict with the contemporary model combined.

If the study of unconscious processes proves to be the key to consciousness, then the unrealised promise of the unconscious that dogged the concept throughout the twentieth century will be finally and decisively realised. Moreover, the realisation of that promise will send as many tremors through scientific and cultural institutions as have been generated by genetics and quantum mechanics.

The brain is a material system, obeying the same laws of physics that influence everything else in the universe. Newton's apple could only take a single path of descent given the forces acting on it at the time; not least of which, of course, was the earth's gravitational field. Similarly, chemical reactions and processes in the brain must conform to a relatively predetermined sequence. If it were possible to take repeated 'snapshots' of the entire brain, then the position of every molecule in any one snapshot should be the logical consequence of molecular events in the preceding snapshot, and so on. If the mind is simply the product of underlying electrochemical activity, then these constraints seem to exclude any possibility of free will.

The seminal work of Benjamin Libet is certainly consistent with this view. His laboratory studies using evoked potential technology demonstrated a half-second delay between the unconscious, electrochemical determinants of behaviour, and the arrival of intentions to behave in awareness.

For some, the idea that human beings are merely sophisticated automata is so abhorrent they have made a valiant effort to rescue 'free will' with the help of quantum physics.

In the quantum realm, what we recognise as the 'common sense' rules that govern the world seem to break down. Things that could not happen with relatively large objects (like cells or people) become possible. For example, electrons can appear to be located at two places at once, or be influenced by what they 'might do' in the future. Such flexibility clearly has explanatory potential with respect to a 'physical' basis of free will. In 1951 the physicist David Bohm wrote:

> … the remarkable point by point analogy between thought processes and quantum processes would suggest that a hypothesis relating these two may well turn out to be fruitful. If such an hypothesis were verified, it would explain in a natural way a great many features of our thinking.

Since 1951 numerous commentators have speculated on the viability of a 'quantum mind'; however, a powerful criticism of this work is that it is simply a reworking of the homunculus solution. Instead of speculating about a special area in the brain where consciousness is generated, scientists now speculate about a special quantum state. Unfortunately, as with the homunculus solution, there is as yet no more evidence linking quantum perturbations and consciousness than there is linking a special area of the brain and consciousness.

We instinctively baulk at the implications of contemporary neuroscience: our behaviour is determined by unconscious processes; the self is epiphe-

nomenal – and a convenient fiction that provides us with post-hoc rationalisations to 'explain' our behaviour; and choice is something that is exercised
in the absence of awareness. In reality, we can make nothing more of ourselves than we already are, and all our major decisions are made a good half-
second before we *think* of them.

In the twenty-first century, Freud has finally landed his third blow. It is
every bit as powerful as he claimed, and threatens to leave the human race in
an ontological swoon; however, this need not be so. Why should the facts
of our existence engender fear and suspicion rather than acceptance and
wonder?

The idea that the self is an illusion only arouses consternation in those
with a deep attachment to an idea of mind in which the conscious ego has a
pivotal or central position. Attachment to this idea is by no means universal.
It is a local phenomenon – associated only with western attitudes and values.
In the east, the illusory nature of self has been recognised for millennia. Curiously, modern neuroscience has reached a virtually identical conclusion to
that reached by Buddhists, who have understood and accepted that the self
is unreal for two and a half thousand years.

Contemporary neuroscience and Buddhism are so close in this respect, it
is possible to accept both paradigms without experiencing any conflict. In an
article titled 'Back to basics' published in 1997, Susan Blackmore – who is at
the same time a psychologist, brain scientist, and practising Buddhist, wrote
the following:

> [The] scientific view is one that gradually reveals how and why we come to
> believe in an illusion. The illusion is that there is somebody in there; that
> consciousness does things; that we as independent, separately existing
> conscious entities run our lives and move our arms and legs. Scientific
> analysis finds no such entity inside the brain, and no need for one.

Blackmore then goes on to sound the registers shared by the scientific and
Buddhist models of mind.

> The Buddha taught, two and a half thousand years ago, that 'actions do exist,
> and also their consequences, but the person that acts does not' and
> Buddhaghosa says, 'Here suffering exists, but no sufferer is found.' There's
> nobody in there … The exciting ideas in science, and my own spiritual
> practice, are both going the same way; the 'no-self' way, if you like; the 'no-
> power-of-consciousness' way. It is a scary way, and a difficult way, but I think
> it is true.

The Buddhist view is that identity (or consciousness) is to the brain what the shape of a wave is to a moving body of water. The wave exists, but only in a very limited sense. Contemporary neuroscience shares this view entirely: the unconscious activity of the total brain produces electrochemical ripples and the shape of those ripples is us. Yes, we exist. But only in a very special sense, and even then, only just.

If it is accepted that the conscious self and the conscious exercise of free will are both illusory, then this must have enormous implications for our moral and ethical codes. To what extent is it right to punish someone for transgressions which, ultimately, are the result of automatic and unconscious brain processes? In reality, a thief steals because he or she can't do anything else. Likewise, a philanthropist is in the same position – giving, because he has to. Social institutions punish the former and reward the latter, assuming the existence of an agent self. But there is no agent self! In an absolute sense, our social institutions may serve much the same function as consciousness – glossing over inconsistencies, providing post-hoc rationalisations, and shoring up an entirely fallacious and misconceived sense of what is right and wrong. In truth, the new model of mind makes concepts such as right and wrong entirely redundant. They simply don't apply.

What, then, is the alternative? Surely we must act as if people are making choices, even if we know the sorry reality?

This may well be true, but even so there would be considerable merit in reminding ourselves of what it really means to be human before expressing contempt or passing judgement. Again, contemporary neuroscience informs a moral sensibility that is oddly eastern. Buddhist and Taoist universes are impersonal. The polarities of yin and yang are reflected in a multiplicity of other dichotomies, among which can be found good and evil. They exist as complementary potentials in a universe that just is. People just are. There is as much point to becoming angry about the behaviour of a thief as there is to becoming angry about the position of a stone. Neither make choices. They occupy their places in the natural order of things.

The philosophical, moral, and scientific implications of recognising the importance of unconscious mental processes are extraordinary. It has taken humanity some time to come to terms with Copernicus and Darwin. Coming to terms with Freud may take even longer, because the third blow demands a complete revision of our understanding of what it means to be human.

Freud never received the Nobel Prize. But, in retrospect, perhaps his one great insight was worthy of such an exalted accolade.

One of the finest works to be written on the dynamic unconscious – as characterised principally by nineteenth-century romantic philosophy and subsequently psychoanalysis – is *The Discovery of the Unconscious* (1970) by Henri Ellenberger. It is a work of great scholarship, providing the reader with an exhaustive survey of key figures and developments in psychodynamic psychiatry up until the Second World War. As we know, the war proved to be a watershed in the history of the unconscious. Thereafter, the disciplines of computer science, cognitive psychology, evolutionary psychology, and neurology all contributed ideas, experimental techniques, and technological advances that helped to establish a new model of the unconscious (albeit a model that revived ideas originally proposed by luminaries such as Huxley, Laycock, Carpenter, and Helmholtz) – the unconscious of automatic reflexes and preconscious processing.

In the final pages of Ellenberger's masterpiece he rightly observes that the old psychodynamic unconscious is incompatible with contemporary science (particularly in the guise of experimental psychology). Clearly, it is impossible to measure libido or quantify the id. No one will ever produce a scaled map of a dreamscape. Subsequently, the dynamic unconscious can never be regarded as a subject suitable for proper scientific study. Indeed, the study of the dynamic unconscious has always been, and will always be, a form of pseudo-science.

Advocates of the dynamic unconscious have been sanguine in the face of this criticism. They have never ceded the superiority of scientific truth above psychological truth, choosing instead to argue that, although it might not be possible to produce a map of the dream world, you can certainly go there – and that too is evidence. Not inferior evidence, but merely evidence of a different kind.

Ellenberger writes:

Actually, we have to deal with two conceptions of reality facing each other, and it would seem that the realm of psychic life can be approached from two sides, both legitimate: either with the accurate technique of measurement, quantification, and experimentation of the research specialist, or with the immediate, nonquantifiable approach of the dynamic psychotherapist.

Ellenberger goes on to suggest that a rapprochement between psychodynamic and scientific psychology might be best achieved by philosophers. He proposes that extended reflection on the nature of psychic reality might reveal what he calls a 'higher synthesis' – a conceptual framework that would do justice to 'the rigorous demands of experimental psychology and to the psychic realities experienced by the explorers of the unconscious'.

Ellenberger does not, however, offer us any further clues as to what this 'higher synthesis' might involve.

When Ellenberger wrote his conclusion, brain-scanning technology was in its infancy. It was possible to do little more than X-ray the brain in order to reveal gross structural abnormalities. Today it is possible to track – and measure – the physical changes that accompany even the most subtle mental operations. Moreover, these mental operations include both conscious and unconscious processes. Already a number of neuroscientists have been bold enough to examine the results of brain-scanning studies, and speculate on where Freudian mechanisms (such as repression) might take place in the brain.

Advances in technology have begun to realise Ellenberger's hope for a marriage between psychodynamic and experimental psychology. The dynamic unconscious and scientific measurement may not be irreconcilable after all.

Yet, however the unconscious is construed – be it hidden intelligence or network of insensible reflexes – one thing is certain: the unconscious has proved to be one of the most robust concepts in psychology. It has survived centuries of scepticism, criticism, and even the odd scandal, to become firmly established at the centre of contemporary neuroscience.

Virtually every aspect of mental life is connected in some way with mental events and processes that occur below the threshold of awareness. Irrespective of the division that separates pre-war, and post-war understanding of the unconscious, the profound importance of unconscious procedures, memories, beliefs, perceptions, knowledge, and emotions is recognised universally. Moreover, for the first time in the history of psychology, scientists are equipped with technologies of sufficient power to undertake a thorough exploration of the unconscious: an exploration that will very probably transform the general conception of what it is to be human. Even basic assumptions about the nature of identity and our ability to make choices will be challenged.

The flame of consciousness, once proudly portrayed as a brilliant and illuminating torch by Enlightenment thinkers, will very likely appear increasingly insignificant as exploration of the unconscious continues. A guttering candle perhaps – or even less.

If the external universe is composed – as astronomers tell us – mostly of dark matter, then it would seem that the internal universe is much the same. We are creatures not of light, but of darkness. A prospect that is unsettling, but alive with extraordinary and exciting possibilities.

Bibliography

Double dates of publication refer to the original publication date and the date of the edition consulted, respectively. The works listed are those of most relevance to the chapters concerned.

1 Depths below depths

De Quincey, T. (1822/1986). *Confessions of an English Opium Eater* (ed. Alethea Hayter, 1971). Penguin: Harmondsworth.

—— (1845/1985). Suspiria de Profundis. In *Confessions of an English Opium Eater and Other Writings.* (ed. Aileen Ward). Carroll & Graff: New York.

—— (1862/1986). *Recollections of the Lakes and the Lake Poets* (ed. David Wright, 1970). Penguin: Harmondsworth.

Ellenberger, Henri F. (1970/1994). *The Discovery of the Unconscious: The History and Evolution of Dynamic Psychiatry.* Fontana: London.

Fancher, R. E. (1990). *Pioneers of Psychology*, 2nd edn. W. W. Norton: New York.

Holmes, R. (1999). *Coleridge: Early Visions*. Flamingo: London.

—— (1999). *Coleridge: Darker Reflections*. Flamingo: London.

Jardine, L. (2000). *Ingenious Pursuits*. Abacus: London.

Lindop, G. (1981/1993). *The Opium Eater: A Life of Thomas De Quincey.* Weidenfeld & Nicolson: London.

MacDonald Ross, G. (1984). *Leibniz*. Oxford University Press: New York.

Murray, J. D. (1988). *A History of Western Psychology*, 2nd edn. Prentice-Hall: New York.

Peirce, C. S. and Jastrow, J. (1884). On small differences of sensation. *Memoirs of the National Academy of Science, 3*, 73–83.

2 Mystery and imagination

Ackroyd, P. (1991). *Dickens*. Minerva: London.

Asimov, I. (1987). *Asimov's New Guide to Science, a Revised Edition*. Penguin: Harmondsworth.

Broughton, R. (1992). *Parapsychology*. Rider: London.

Buranelli, V. (1975). *The Wizard from Vienna*. Coward, McCann, and Geoghegan: New York.

Carey, J. (ed.) (1995). *The Faber Book of Science*. Faber and Faber: London.

Ellenberger, Henri F. (1970/1994). *The Discovery of the Unconscious: The History and Evolution of Dynamic Psychiatry*. Fontana: London.

Encyclopaedia Britannica (1999). Standard edition on CD-ROM.

Fancher, R. E. (1990). *Pioneers of Psychology*, 2nd edn. W. W. Norton: New York.

Forrest, D. (2000). *Hypnotism: A History*. Penguin: Harmondsworth.

Kahl, R. (1971). *Selected Writings of Hermann von Helmholtz*. Wesleyan University Press: Middletown, CT.

Miller, J. (1997). Going Unconscious. In Robert B. Silvers (ed.), *Hidden Histories of Science*. Granta: London.

3 The philosophy teacher

Drinka, G. F. (1984). *The Birth of Neurosis: Myth, Malady and the Victorians*. Simon & Schuster: New York.

Ellenberger, Henri F. (1970/1994). *The Discovery of the Unconscious: The History and Evolution of Dynamic Psychiatry*. Fontana: London.

Mayo, E. (1948). *Some Notes on the Psychology of Pierre Janet*. Harvard University Press: Cambridge, MA.

Myers, F. W. H. (1886). On Telepathic Hypnotism, and its relation to other forms of hypnotic suggestion. *Journal of the Society for Psychical Research*, 4, 127–88.

Reed, G. F. (1987). Pierre Janet. In Richard L. Gregory (ed.), *The Oxford Companion to The Mind*. Oxford University Press: Oxford.

4 The icon

Appignanesi, L. and Forrester, J. (1992/2000). *Freud's Women*. Penguin: Harmondsworth.

Breuer, J. and Freud, S. (1893–95/1991). *Studies on Hysteria*, Penguin Freud Library Vol. 3 (trans. James and Alix Strachey, ed. James and Alix Strachey and Angela Richards). Penguin: Harmondsworth.

Cottrell, L. (1953/1992). *The Bull of Minos*. Efstathiadis: Athens.

Freud, S. (1900/1991). *The Interpretation of Dreams*, Penguin Freud Library Vol. 4 (trans. James Strachey, ed. James Strachey with Angela Richards and assisted by Alan Tyson). Penguin: Harmondsworth

—— (1909/1990). Notes upon a case of obsessional neurosis (the 'rat man'). In *Case histories II*, Penguin Freud Library Vol. 9. (trans. and ed. James Strachey, comp. and ed. by Angela Richards). Penguin: Harmondsworth

—— (1913[1912–13]/1990).Totem and Taboo. In *The Origins of Religion*,

Penguin Freud Library Vol. 13 (trans. and ed. James Strachey (and later ed. Albert Dickson)). Penguin: Harmondsworth

—— (1915/1991). The unconscious. In *Metapsychology: The Theory of Psychoanalysis*, Penguin Freud Library Vol. 11 (trans. James Strachey, comp. and ed. Angela Richards). Penguin: Harmondsworth

—— (1917[1916–17]/1987). Lecture 18, Fixation to traumas – the unconscious. In *Introductory Lectures on Psychoanalysis*, Penguin Freud Library Vol. 1. (trans. James Strachey, ed. James Strachey and Angela Richards). Penguin: Harmondsworth.

Gay. P. (1988) *Freud: A Life for Our Time*. J. M. Dent & Sons: London.

Showalter, E. (1987). *The Female Malady: Women, Madness and English Culture, 1830–1980*. Virago: London.

Tallis, F. (1998). *Changing Minds*. Cassell: London.

5 Darkness rising

Brown, J. A. C. (1961/1994). *Freud and the Post-Freudians*. Penguin: Harmondsworth.

Ellenberger, Henri F. (1970/1994). *The Discovery of the Unconscious: The History and Evolution of Dynamic Psychiatry*. Fontana: London.

Encyclopaedia Britannica (1999). Standard edition on CD-ROM.

Fordham, F. (1953/1972). *An Introduction to Jung's Psychology*. Penguin: Harmondsworth.

Freud, S. and Jung, C. (1974/1991). *The Freud/Jung Letters* (ed. William McGuire). Penguin: Harmondsworth.

Gay. P. (1988) *Freud: A Life for Our Time*. J. M. Dent & Sons: London.

Harman, A., Miller, A. and Mellers, W. (1962/1971). *Man and His Music*. Barrie & Jenkins: London.

Jung, C. (1962/1982). *Memories, Dreams, Reflections*. Collins Fount Paperbacks: Glasgow.

McClung B. (1997). The Saga of *Lady in the Dark*. National Theatre programme for the 1997 production of *Lady in the Dark* (opening The Lyttelton Theatre, 11 March 1997).

Storr, A. (ed.) (1986). *Jung: Selected Writings*. HarperCollins: Glasgow.

Tallis, F. (1998). *Changing Minds*. Cassell: London.

6 A new vocabulary

Bargh, J. and Pietromonaco, R. (1982). Automatic information processing and social perception: the influence of trait information presented outside of conscious awareness on impression formation. *Journal of Personality and Social Psychology*, 43, 437–49.

Broadbent, D. (1958). *Perception and Communication*. Pergamon: Oxford.

Cherry, E. C. (1953). Some experiments on the recognition of speech with one and two ears. *Journal of the Acoustical Society of America*, 25, 975–79.

Dixon, N. (1981). *Preconscious Processing*. John Wiley & Sons: Chichester.

Eysenck, M. (1984). *A Handbook of Cognitive Psychology*. LEA: London.

Kihlstrom, J. (1987). The Cognitive Unconscious. *Science*, 237, 1445–52.

Kunst-Wilson, W. and Zajonc, R. (1980). Affective discrimination of stimuli that cannot be recognized. *Science*, 207, 557–58.

Lazarus, R. and McCleary, R. (1951). Autonomic discrimination without awareness: a study of subception. *Psychological Review*, 58, 113–22.

Lewicki, P., Hill, T. and Czyzewska, M. (1992). Nonconscious acquisition of information. *American Psychologist*, 47 (6), 796–801.

Miller, G. (1956). The magical number seven, plus or minus two: some limits on our capacity for processing information. *Psychological Review*, 63, 81–87.

Nisbett, R. and Wilson, T. (1977). Telling more than we can know: verbal reports on mental processes. *Psychological Review*, 84, 231–59.

Postman, L., Bruner, J. and McGinnies, E. (1948). Personal values as selective factors in perception. *Journal of Abnormal Psychology*, 43, 142–54.

Reed, G. F. (1987). Time gap experience. In Richard L. Gregory (ed.), *The Oxford Companion to The Mind*. Oxford University Press: Oxford.

Rheingold, H. (1985). *Tools for Thought: The People and Ideas of the Next Computer Revolution*. Simon & Schuster: New York.

7 The unconscious brain

Baddeley, A. (1990). *Human Memory: Theory and Practice*. LEA: Hove and London.

Burke, J. (2000). Harsh words can deform children's brains for life. *Observer*, 31 December.

Carter R. with Frith C. (2000). *Mapping the Mind*. Phoenix: London.

Crick, F. and Koch, C. (1992). The problem of consciousness. In J. Rennie (ed.), *Scientific American* special issue: *Mysteries of the Mind*, 19–26.

Damasio, A. (2000). *The Feeling of What Happens: Body and Emotion in the Making of Consciousness*. Vintage: London.

Deecke, L., Grozinger, B. and Kornhuber, H. (1976). Voluntary finger movement in man: cerebral potentials and theory. *Biological Cybernetics*, 23, 99–119.

Fancher, R. E. (1990) *Pioneers of Psychology*, 2nd edn. W. W. Norton: New York.

Gazzaniga, M. (1967). The split brain in man. *Scientific American*, 217, 24–29.

—— (1992). *Nature's Mind: The Biological Roots of Thinking, Emotions, Sexuality, Language and Intelligence*. Penguin: Harmondsworth.

Le Doux, J. (1985). Brain, mind, and language. In D. Oakley (ed.), *Brain and Mind*. Methuen: London.

Lezak, M. (1983). *Neuropsychological Assessment*, 2nd edn. Oxford University Press: Oxford.

Libet, B. (1985). Unconscious cerebral initiative and the role of conscious will in voluntary action. *Brain and Behavoural Sciences*, 8, 529–66.

—— (1991). Conscious vs. neural time. *Nature*, 352, 27.

Libet, B., Alberts, W., Wright Jr., E., Delatytre, L., Levin, G. and Feinstein, B. (1964). Production of threshold levels of conscious sensation by electrical stimulation of human somatosensory cortex. *Journal of Neurophysiology*, 27, 546–78.

Libet, B., Alberts, W., Wright, E. and Feinstein, B. (1967). Responses of human somatosensory cortex stimuli below the threshold for conscious sensation, *Science*, 158, 1597–1600.

McCone, J. (1999). *Going Inside*. Faber and Faber: London.

Meno, D., Owen, A., Williams, E., Minhas, P., Allen, C., Boniface, S., Oickard, J., Kendall, I.,Downer, S., Clark, J., Carpenter, T. and Antound, N. (1998). Cortical processing in persistent vegetative state. *The Lancet*, 352, 800.

Parkin, A. (1996). The alien hand. In P. Halligan and J. Marshall (eds.), *Method in Madness: Case Studies in Cognitive Neuropsychiatry*. Psychology Press: Hove.

Sperry, R. (1968). Hemisphere disconnection and unity in conscious awareness. *American Psychologist*, 23, 723–33.

Thompson, R. (1975). *Introduction to Physiological Psychology*. Harper & Row: New York.

Trevarthen, C. (1987). Split brain and the mind. In Richard. L. Gregory (ed.), *The Oxford Companion to The Mind*. Oxford University Press: Oxford.

Weiskrantz, L. (1986). *Blindsight: A Case Study and its Implications*. Clarendon Press: Oxford.

Weiskrantz, L., Warrington, E., Sanders, M. and Marshall, J. (1974). Visual capacity in the hemianopic field following a restricted occipital ablation. *Brain*, 97, 709–28.

8 Darwin in the dark

Badcock, C. (1995). *PsychoDarwinism: The New Synthesis of Darwin and Freud*. Flamingo: London.

Caplan, A. (ed.) (1978). *The Sociobiology Debate: Readings on the Ethical and*

Scientific Issues Concerning Sociobiology (with a foreword by Edward
  O. Wilson). Harper & Row: London.
Carter R. with Frith C. (1998/2000). *Mapping the Mind*. Phoenix: London.
Damasio, A. (1996). *Descartes' Error*. Macmillan: London.
Dawkins, R. (1976). *The Selfish Gene*. Oxford University Press: Oxford.
—— (1988). *The Blind Watchmaker*. Penguin: Harmondsworth.
Edelman, G. (1987). *Neural Darwinism: The Theory of Neuronal Group Selection*.
  Basic Books: New York.
Ekman, P. (1986). *Telling Lies*. Berkley Books: New York.
Ekman, P. and Friesen. W. (1974). Detecting deception from the body or
  face. *Journal of Personality and Social Psychology,* 29, 288–98.
Gur, R. and Sackheim, H. (1979). Self-deception: a concept in search of a
  phenomenon. *Journal of Personality and Social Psychology*, 37, 147–69.
Le Doux, J. (1994). Emotion, memory and the brain. In J. Rennie (ed.),
  Scientific American special issue: *Mysteries of the Mind*, 68–75.
—— (1996). *The Emotional Brain*. Simon & Schuster: New York.
MacLean, P. (1976). Sensory and perceptive factors in emotional function of
  the triune brain. In R. Grenell and S. Gabay (eds.), *Biological Foundations of
  Psychiatry*. Raven: New York.
Meadows, J. (ed.) (1987). *The History of Scientific Discovery*. Phaidon: Oxford.
Mogg, K. and Bradley, B. (1999). Selective attention and anxiety: a cognitive-
  motivational perspective. In T. Dalgleish and M. Power (eds.), *Handbook of
  Cognition and Emotion*. John Wiley & Sons: Chichester.
Öhman, A. (1999). Distinguishing unconscious from conscious emotional
  processes: methodological considerations and theoretical implications. In
  T. Dalgleish and M. Power (eds.), *Handbook of Cognition and Emotion*. John
  Wiley & Sons: Chichester.
Ridley, M. (1994). *The Red Queen: Sex and the Evolution of Human Nature*.
  Penguin: Harmondsworth.
Sackheim, H. (1983). Self-deception, self-esteem, and depression: the
  adaptive value of lying to oneself. In J. Masling (ed.), *Empirical Studies of
  Psychoanalytic Theories*, Vol. 1, Analytic Press: Hillsdale, NJ, 101–57.
Sagan, C. (1981). *Cosmos*. Futura: London.
Seligman, M. (1971). Phobias and preparedness. *Behaviour Therapy*, 2, 307–20.
Trivers, R. (1985). *Social Evolution*. Benjamin/Cummings: Menlo Park, CA.
Wright, R. (1996). *The Moral Animal*. Abacus: London.

9 The uses of darkness

Anastasi, A. (1954/1982). *Psychological Testing*, 5th edn. Collier Macmillan:
  London.

Andrade, J. (in press). Learning during sedation, anaesthesia, and surgery. In M. M. Ghoneim (ed.). *Awareness During Anesthesia*. Butterworth Heinemann: Woburn, MA.

Beech, H. R. (1959). An experimental investigation of sexual symbolism in anorexia nervosa employing a subliminal stimulation technique: a prelimininary report. *Psychosomatic Medicine*, 21, 277–80.

Dixon, N. (1981). *Preconscious Processing*. John Wiley & Sons: Chichester.

Green, C. and McCreery, C. (1994). *Lucid Dreaming: The Paradox of Consciousness During Sleep*. Routledge: London.

Hardaway, R. (1990). Subliminally activated symbiotic fantasies: facts and artifacts. *Psychological Bulletin*, 107 (2), 177–95.

Hervey de Saint-Denis, M.-J. (1867). *Les Rêves et les moyens de les diriger*. Amyot: Paris.

Howe, J., Ashworth, P., Blackmore, S., Blagrove, M., Henley, S. and Underwood, G. (undated report). *Subliminal Messages*. British Psychological Society: Leicester.

Kelzer, K. (1987). *The Sun and the Shadow: My Experiment with Lucid Dreaming*. ARE Press: Virginia Beach, VA.

Kihlstrom, J. (1987). The cognitive unconscious. *Science*, 237, 1445–52.

Loftus, E. and Klinger, M. (1992). Is the unconscious smart or dumb? *American Psychologist*, 47, (6), 761–65.

Poeztl, O. (1917/1960). The relationship between experimentally induced dream images and indirect vision. Monograph No. 7, *Psychological Issues*, 2, 41–120.

Rorschach, H. (1921/1942). *Psychodiagnostics: A Diagnostic Test Based on Perception* (trans. P. Lemkau and B. Kronenburg). Huber: Berne.

Silverman, L. (1966). A technique for the study of psychodynamic relationships: the effects of subliminally presented aggressive stimuli on the production of pathological thinking in a schizophrenic population. *Journal of Consulting Psychology*, 30, 103–11.

—— (1976). Psychoanalytic theory: 'the reports of my death are greatly exaggerated'. *American Psychologist*, 31, 621–37.

Silverman, L., Bronstein, A. and Mendelsohn, E. (1976). The further use of the subliminal psychodynamic activation method for the experimental study of the clinical theory of psychoanalysis: on the specificity of relationships between manifest psychopathology and unconscious conflict. *Psychotherapy: Theory, Research, and Practice*, 13 (1), 2–16.

Silverman, L., Frank, S. and Dachinger, P. (1974). A psychoanalytic interpretation of the effectiveness of systematic desensitization: experimental data bearing on the role of merging fantasies. *Journal of*

Abnormal Psychology, 83, 313–18.

Silverman, L., Martin, A., Ungaro, R. and Mendelsohn, E. (1978). Simple research paradigm for demonstrating subliminal psychodynamic activation: effects of Oedipal stimuli on dart-throwing accuracy in college males. *Journal of Abnormal Psychology*, 87, 341–57.

Stevens, A. (1996). *Private Myths: Dreams and Dreaming*. Penguin: Harmondsworth.

Storr, A. (1976). *The Dynamics of Creation*. Penguin: Harmondsworth.

Tyrer, P., Lee, I. and Horn, S. (1978). Treatment of agoraphobia by subliminal and supraliminal exposure to phobic cine film. *The Lancet*, 18, 358–60.

Vance, J. et al. v. *Judas Priest et al.* (24 August 1990) No. 86–5844, 2nd District Court of Nevada.

Weinberger, J. (1992). Validating and demystifying subliminal psychodynamic activation. In R. Bornstein and T. Pittman (eds.), *Perception Without Awareness*. Guilford Press: London.

Whitworth, D. (2000). Bush forced to withdraw 'rat' advert. *The Times*, 13 September.

10 The third blow

Blackmore, S. (1997). Back to basics. *Journal of the Society for Psychical Research*. 61, (846), 333–35.

Bohm, D. (1951). *Quantum Theory*. Prentice-Hall: New York.

Damasio, A. (2000). *The Feeling of What Happens: Body and Emotion in the Making of Consciousness*. Vintage: London.

Dennett, D. (1993). *Consciousness Explained*. Penguin: Harmondsworth.

Ellenberger, Henri F. (1970/1994) *The Discovery of the Unconscious: The History and Evolution of Dynamic Psychiatry*. Fontana: London.

Freud, S. (1917[1916–17]/1987). *Introductory Lectures on Psychoanalysis*, The Penguin Freud Library Vol. 1 (trans. James Strachey, ed. James Strachey and Angela Richards). Penguin: Harmondsworth.

Gay. P. (1988). *Freud: A Life for Our Time*. J. M. Dent & Sons: London.

Watts, A. (1957/1990). *The Way of Zen*. Penguin Arkana: Harmondsworth.

Zohar, D. (1991). *The Quantum Self*. Flamingo: London.

Index